MANGROVE MAGIC

MANGROVE MAGIC

Game of Clones

Ghazally Ismail

To order additional copies of this book, contact:
Xlibris
NZ TFN: 0800 008 756 (Toll Free inside the NZ)
NZ Local: 9-801 1905 (+64 9801 1905 from outside New Zealand)
www.Xlibris.co.nz
Orders@Xlibris.co.nz
846446

CONTENTS

Dedication...vii

1. Introduction..1
2. Unique Mangroves...9
3. Exclusive Residence 25
4. Visitors Welcomed 37
5. Transit For World Travellers...55
6. Getting About 73
7. Out Of Harm's Way 89
8. Got You Covered 109
9. Staying Connected...127
10. Good With Food 147
11. Making More Of The Same..179
12. Game Changer In Climate Change.....................................195

About the Author.. 209

DEDICATION

A scientist has a profound obligation to society and humanity. A cause and commitment one must endeavour to fulfil even past working life. I have chosen to spend my retirement disseminating factual information about the basic sciences underlying the health of our environment. I feel this deep responsibility to continue educating the policy-makers and public at large on issues related to biodiversity and environmental conservation. I intend to do this in a manner that can be understood by lay-people and youths of tomorrow. A very special kind of communication that was made easier by being around special people in my life.

I am forever indebted to my wife, Lesley, who had been by my side throughout my career as an academic and researcher. Famished and mentally drained, I was always guaranteed of her support during my writing this book. Lunch sandwiches, coffee and ice-cream for supper would always be delivered in perfect timing to the library at home. What I have to write in here, my children Azizah, Salina, Jeffry and Zachary have heard them all from me in their childhood days in Borneo. That helped made completing this project a breeze.

I wish to specially dedicate this book to my grandchildren, Alyssa, Nate, Jassin, Baxter and Caspar. May they be endowed with the education, awareness and convictions leading to an assurance that the future of our natural ecosystems and biodiversity is in good hands.

1

Introduction

The mangroves are mysterious places. It is a unique habitat where many plant and animal species have adapted to exist and thrive. Every close scrutiny of life there reveals the mind-blowing biological and behavioural features, which have come into play in their ability to survive and prosper for millions of years under inhospitable conditions. Every species is a victor of evolution and adaptation in its own right. They all have been champions in a perplexing game of clones. They are groups of individual organisms genetically derived from the original species, which could have been far from perfect to survive in the harsh environment of the mangroves. They are cells or organisms having a survival advantage over those from which they are derived. They have evolved and specifically adapted to live in the mangroves. Over millions of years, the inferior genes of their progenitors were eliminated, and the superior genes were selected through evolutionary processes and adaptations. The victorious and surviving species inhabiting the mangroves have continued to enthral us till today.

Mangroves are invariably found stretching along water bodies throughout the tropics such as the coasts, estuaries, rivers and lakes. In Malaysia, mangrove forests occur primarily along the west coast of Peninsular Malaysia, the estuaries of rivers of Sarawak, and the coasts and islands of Sabah. The latest census revealed less than 2 per cent of the total land area in Malaysia comprised mangroves. As a young boy, I remember admiring the pristine condition of mangrove forests stretching as far as my eyes could see. They were primarily untouched by human hands at that time. No one was interested in building their dwellings and living next to a mangrove forest despite the lush green landscape it presented. Humid smelly air that seemed to be buzzing with mosquitoes just didn't appeal to most people.

But I have always been intrigued with the mangroves. I was filled with amazement about this unique ecosystem even growing up as a child. There were mysteries everywhere. Big and small questions popped up in my head as I wondered about how nature actually worked here. Why did trees have long green fruits hanging from their branches? There were alien-looking fish with a bulging pair of eyes that were able to walk on land. Why did it smell disgusting all the time? What were those millions of needle-like things coming out of the mud under the trees pointing straight up to the sky?

My natural interest towards mangroves waned a little because of their physical distances in later years. I left my rural surroundings of Jelawat Village at the age of twelve to spend seven years in a boarding school in Kuala Kangsar, a quaint urban setting on the west coast of Peninsular Malaysia. It was a stark difference from the rural setting to which I was accustomed. There were no mangrove forests within the vicinity that I could explore and enjoy. But during my time at Malay College, Kuala Kangsar, I found my fondness at drawing rekindled. I thoroughly enjoyed my art classes under the tutelage of an art teacher, Syed Bakar. I was immensely inspired by his ideas and personality that seemed to go off tangent to other teachers. Despite my love for arts and creativity, I ended up studying science in school. Malaysia, as a newly independent country, required more doctors, engineers, and scientists, so they told me. I completed a science degree in microbiology at the University of Otago, New Zealand. My student days abroad took me away from my childhood enclaves. There were no muddy flats and tangling mass of mangrove roots to explore. It was during my undergraduate days in a botany course at Otago University when I started reading about the uniqueness of the mangrove ecosystems. I found many of the seemingly simple and naive questions I used to ask as a child actually had complex scientific answers. They were far from obvious. In fact, as a biology student, I found some of the explanations to my childhood curiosities most enlightening. They made me understand the scheme of things and how evolution had come to the positive side of nature. Indeed, millions of years of evolution and adaptations have made mangroves the highly productive and successful ecosystem we have around us today.

After my doctoral degree from Indiana University-Purdue University Indianapolis United States, I returned to Malaysia to serve as a young university lecturer in Kuala Lumpur. I began to feel a renewed interest in mangroves. My fascination for this unique ecosystem was further ignited by the opportunities to join my fellow biology lecturers undertaking field research in the mangroves. I continued to ask probing questions and learned from them. Once a year in the mid-1970s, my young family would make a road trip to celebrate a religious holiday, Hari Raya Aidilfitri, celebrated by Muslims, which marks the end of the Islamic holy month Ramadan. It was a long drive from Malaysia's capital, Kuala Lumpur, to my small village, Jelawat, on the east coast of Peninsular Malaysia. I was then working at the newly established Universiti Kebangsaan Malaysia (UKM) at its temporary campus in Jalan Pantai, Kuala Lumpur. For a Malay, during festive holidays like Aidilfitri, it was a custom to be with your parents in the house in which you grew up as a child. Annually, this would be the road trip much looked forward to by everyone, especially those advancing their careers in the bustling cities away from the rural kampungs in which they were born and raised. My wife, Lesley, would plan a picnic stop to break the ten-hour journey. It was more out of a necessity. The stop was not just to rest my stiff neck from driving but also to cease the usual moaning and whining of our two young children cooped up in the back seats during the long journey. Much to the relief of everyone, I'd stop at a nice open shady spot to have our lunch. But Lesley would without fail warn me beforehand not to stop next to a mangrove. She just couldn't enjoy a picnic next a smelly mangrove. Of course, it would neither be ideal nor fair to enjoy the picnic she had spent days preparing for the occasion. Lesley would usually cook *nasi lemak*, a favourite Malaysian meal comprising basmati rice specially cooked in coconut milk with a blade or two of the pandanus leaf thrown in for the aroma. No one in their right mind would want to eat nasi lemak with the off-putting odour hovering in the air, no matter how scenic the picnic spot would be. The children, however, would express their disappointment at not having the chance to sink their feet in the muddy black sand of the mangroves and chase the mudskippers bopping up and down hurriedly into their holes. I would proudly tell them of my own childhood experiences stomping around in the mangroves. Drawing their attention to the hundreds of aerial roots of the mangrove trees pointing up to the clear blue sky above, I would tell them of the time when Jelawat children, including their father, used to cut those tapering pointed roots and pretend they were our *keris*, a traditional dagger-like weapon of my ancestral Malays. We would be donning the make-believe keris by our waist, walking around like warriors in front of the girls in the village.

The spectacle of how a mangrove ecosystem established itself had always fascinated me. The persistence and dynamics of its formation was phenomenal. The ebbing tide slowly but surely nibbled away at the land. The blackish sandy banks bordering the mangroves gradually eroded away, discharging tons of silt and mud into the sea. Additionally, rivers and waterways from the mainland continually deposited many tons of naturally decomposing materials and plant debris into the sea. Over the years, the land expanded seaward, spreading far and wide into the shallow coastal waters. This resulted in a build-up of slimy muddy basements further into the seafront, creating habitats favourable for the growth and establishment of specialised plant communities. Salt-tolerant mangrove species grew well on the newly formed soft muddy substratum, gradually pushing the boundary of the green belt towards the sea. Every year, more and more of the ocean would be reclaimed. That was the genesis of most mangrove forests in the tropics.

Naturalists who came to the tropics were immediately captivated by the unique lures of the mangrove forests. The mangroves posed such an irresistible magnet. Visitors were instantly mesmerised by an inimitable ecosystem at the edge of the sea, which harboured a variety of fascinatingly weird creatures from the land and the sea. The mangroves boasted of fish that could walk on land, fish that could shoot down insects with powerful squirts of water, and fireflies that lit up rows of trees into glowing Christmas spectacles at night.

But in modern times, these same mangrove forests are in peril. They have been plundered for various reasons. The pressure was on for the massive clearance of mangrove forests to make way for human activities. Large tracts of mangroves were drained for land reclamation, highly sought after for the construction of housing estates, factories, hotels, resorts, and aquaculture ponds. This unique natural ecosystem has been disappearing at unprecedented rates throughout the tropics, including Malaysia. I became anxious if the mangroves would totally disappear before my childhood wonderment about the animals and trees found in this special environment ever got answered. I decided to embark on a journey of my own. Throughout my working life, I made it my educational pursuit to explore the mangroves and their biological components at every opportunity I had. I read widely on the workings and sciences that made mangroves evolve. Today, I can cautiously say most of my curiosities have been met with some exciting explanations. My childhood thirst for knowledge quenched to satiation. I now see mangroves as a tropical ecosystem upending every model of success that has emerged through time. It is an epic story of how biological evolution and adaptation have created the most fascinating and productive ecosystem on our ever-competitive and fragile planet Earth. Mangrove forests are today the best place to tell a compelling story of survival of the fittest for both plant and animal species.

Mangrove flora

Mangrove fauna

I have seen and enjoyed learning about mangroves in variously interesting locations throughout Malaysia. My obsessive tendency to scientifically understand this unique tropical ecosystem has been most rewarding to me as an educator. Dare I say I'm majorly out of the dark now about the many mangrove magic with which I was grappling to make sense as a child. It would be a shame not to share some of the fascinating facts about the myriad species that make up the mangroves. This book includes some of that story. Indeed, everyone ought to know more about this wonderful and important ecosystem on planet Earth that desperately needs our protection.

2

Unique Mangroves

The day was fine with a clear blue sky above. We arrived around mid-morning at Kuala Selangor Nature Park, a large mangrove sanctuary of approximately two hundred hectares established in 1987. Starting with a humble beginning, today it has become an inspiring showcase for the conservation of mangrove ecosystems in the region. Immediately, I was beginning to be very impressed with what I saw. I learned that the area was initially targeted for development to meet the constant demands for human settlements and business districts in this area, a rural enclave only an hour's drive from the capital city, Kuala Lumpur. In the early 1980s, a colossal and ambitious land reclamation project was planned. Luckily, clear and responsible ecological sense prevailed. The plan was aborted. Today, it stands proud as a natural monument that exemplifies Malaysia's commitment to mangrove and wetland conservation.

I immediately noticed the many 'needles' pointing up to the blue sky from the shimmering waters of the mudflats, a throwback to my childhood in Jelawat. But this time, I no longer viewed in admiration the pointed shape of those aerial roots resembling the Malay keris. I still looked at them in awe but from a different perspective. In the comic world, I love reading to my grandchildren about superheroes. What I like most about superheroes is they *never die. Whether humanoid, blue ape, green hulk, gorilla, when the planet is on fire or* asteroids just struck planet Earth, one thing my grandchildren is rest assured, their superheroes will always survive. They might have come close to death, but they somehow come back. That is in the comic world. But in the plant kingdom, there are 'superheroes' that just know how to overcome whatever calamities and adversities occur. They are survivors. These are the mangrove plants I was standing amongst. With their roots submerged in water, mangrove trees thrive and continue to grow and flourish. In the hot, muddy, salty conditions they are in, most plants would have quickly perished. But mangroves have been thriving under such inhospitable conditions for millions of years. Against all odds, mangroves, like superheroes, will always survive! How did they succeed?

Specialised aerial roots called pneumatophores allow the mangroves to breathe in habitats that are majorly waterlogged, and they are growing in oxygen-poor muddy substratum.

Mangrove plants cheat death through a series of impressive adaptations. The salty environment of the intertidal zone is harsh and prohibitive for most plants to flourish. But the mangrove is uniquely adapted for these conditions. To help them survive in salty conditions, they have evolved a filtration system that excludes much of the salt from their tissues. These adaptations are remarkably so successful that some mangroves are able to grow in soils that reach salinities up to 75 parts per thousand (ppt), twice the salinity of ocean water. Some species are able to actively eliminate salt from their tissues, hence known as salt secretors. They employ special glands in their leaves capable of adsorbing excess salts from inside and excreting directly into the environment. White salt crystals on the leaf surfaces of a few mangrove species can be seen with our naked eyes if you care to look closely. Other plant species simply block the entry of salts. Non-secretor species from the genera *Rhizophora* and *Bruguiera* possess a barrier that can almost completely exclude salt from entering their vascular system. Such barriers can be so efficient that over 90 per cent of the salt from seawater can be excluded. This barrier acts against osmosis where water moves from areas of low salt concentration to areas of high concentration. Without such clever mechanisms, the salty ocean water would have sucked the mangroves dry. Other mangrove species have a range of adaptations for coping with their muddy and salty environment. The salty intertidal soils left scarce freshwater needed for the physiological processes to occur within the plant cells. Here, again, the mangrove species have evolved clever means in solving this acute shortage of fresh water in their cells. They simply keep a tight rein on the amount of water lost through their leaves. This is achieved by regulating the opening of their stomata, constricting the size of aperture to prevent excessive loss of water. This way, the rate of exchange of carbon dioxide and water during photosynthesis is effectively reduced.

To save water, the plant also takes ingenious steps to reduce its rate of evaporation. This is adeptly done by constantly changing the orientation of their leaves to avoid receiving direct rays of the harsh midday sun. I nodded in full admiration how these plants adapt accordingly. I stopped asking myself why leaves of some plant species here appeared to stand almost vertically erect around noon time. As the day gets cooler, the leaves would increasingly lie almost flat, facing up towards the sky again, a behavioural adaptation that will further lower the rate of transpiration and retain water in their cells. Truly amazing.

To the uninitiated visitors, their untrained eyes will see all mangroves as being the same with respect to tree composition. But experts have documented at least twenty-two species of mangrove trees found in Malaysia. It didn't take long for me to recognise the iconic *Rhizophora* trees from other mangrove species. They are characterised by complex stilt-like roots radiating from the main trunk, spreading outwards and plunging into the mud in the form of arches. These tangled networks of arches form impenetrable barriers, which not only provide strong support for the trees but also trap the ever-shifting mud, debris, and fine silt brought in by the daily tides.

The three dominant mangrove species known to the locals are *bakau kurap*, *bakau minyak*, and *api-api putih*. Each species can be easily identified through different characteristics of their leaves, flowers, fruits, or roots.

The unmistakable bakau kurap is scientifically referred to as *Rhizophora mucronata*. It grows better on sandy and firmer substrates than other mangrove species. But it still can tolerate the onslaughts of salty seawater as tidal waves rush in assuredly twice a day. I was intrigued by the acute observations of the locals in naming these mangrove trees. *Kurap* is the Malay name for ringworm, a type of skin disease found in humans and domestic animals such as cats and dogs. It is caused by a fungal infection hardly considered a serious affliction. Ringworm only causes itching of the skin, making it appear scaly, pimply, and reddish, especially on hot days. These characteristic manifestations of ringworm could have inspired the locals to equate *Rhizophora mucronata* fruit that exhibits scaly and pimply textures resembling kurap.

The bakau minyak is scientifically known as *Rhizophora apiculata*. It belongs to the same family Rhizophoraceae as bakau kurap. The species is found growing on muddy soils normally inundated by high tide but may also do well in sandy areas amongst *R. mucronata*. From afar, a trained eye may be able to differentiate the lighter colouration of its stems and branches compared to bakau kurap. On closer examination, bakau minyak has dark-grey or dark-brown bark. I find it difficult to see any difference in the overall architecture of their root systems. Both possess complex networks of prop roots and stilt roots coming from higher and lower parts of the trees respectively.

A massive tangle of stilt roots growing out of the trunks of *Rhizophora* species are yet another adaptation allowing the exchange of gas in oxygen-poor mud and sediment.

Api-api putih is scientifically known as *Avicennia alba*, which belongs to the family Acanthaceae. To date, I have yet to find a believable explanation as to how *Rhizophora apiculata* bakau minyak got its name. But determining how *Avicennia alba* got its name, api-api putih is quite straightforward. In many mangrove forests, the occurrence of this species can be spotted from a distance by the white colours of their leaves. Their narrow and pointed leaves, no longer than twelve centimetres, are shiny dark green above and whitish colouration on the underside. *Alba* is Latin for 'white', which translates to *putih* in Malay. The locals must have noticed the stretch of white demarcation from a distance when these leaves are consistently flipped over by the sea breeze during the day and the land breeze just before nightfall. I am sure this led to its Malay name api-api putih. Unlike bakau kurap and bakau minyak, api-api putih has small, slender roots commonly seen protruding from the mud invariably tapering upwards like pencils. They are particularly encountered in newly advancing mud banks facing the sea or along riverbanks and lakesides. Because of this habit, *Avicennia alba* are often considered pioneering species in the formation of mangrove forests.

We ventured deeper into the mangrove forests. Soon enough, I was beginning to get frustrated by the constant zigzagging and hopping around I had to do. Avoiding the obstacles presented by the complex root systems and soft sandy mud of the mangroves was not easy. While treading my feet carefully amongst those roots, I tried to recall the well-known distinguishing features of the various species types of mangroves found in Malaysia. Their root systems are reliable indicators of the different genus of these mangroves. Anyone who had gone through fundamental biology in high school would be able to name two primary functions of plant roots. Firstly, roots are supposed to be growing deep downwards and horizontally outwards in every direction. This way, roots can provide stability to the plant by securely anchoring the entire plant from being blown by the wind or a storm. Secondly, roots take up water and inorganic nutrients needed for growth and survival of the above-ground shoots. Never was any mention of roots having a third major function in breathing. Mangrove roots perform all these three functions—providing stability, uptake of nutrients, and breathing oxygen from the surrounding air. To my mind, the cleverest mangrove trick was to further evolve and adapt their roots to perform this third function besides providing support and drawing in nutrients, to make their roots do multitasking jobs.

I find it most remarkable that besides spreading over a wide area to help stabilise the tree in soft sand and muddy soils, mangrove roots help the plants breathe as well. Air is scarce in the waterlogged soil of the mangroves. To overcome this danger of drowning, the roots emerge out of the ground to become specialised aerial roots called pneumatophores. Oxygen enters the plant cells and tissues through thousands of breathing pores called lenticels. But this only happens when the pneumatophores are exposed during low tide. The lenticels are closed tightly during high tide to prevent the plants drowning. Millions of years of evolution have created an alternative form of roots that can allow mangroves to survive under inhospitable conditions, an out-of-the-ordinary departure from other plant roots that are in other terrestrial plant kingdoms. Pneumatophores with lenticels, which are morphologically and functionally different, have indeed made these plants to be superheroes and survive. I felt a little guilty for my ignorance as a child for harvesting these pointed pneumatophores and pretending they were our keris, the curvy pointed weapon of Malay warriors of our past. The *Rhizophora* has stilt roots, which arch out from both the lower and upper portions of their stems. On closer examination, the plant roots are distinctly characterised by curls and loops, which spread below and above ground at random intervals. They look extremely complex and haphazard. The root systems of other mangrove plant genera *Sonneratia* and *Bruguiera*, however, are quite predictable. They all have solid buttress roots found growing on drier ground farther away from the sea edge, often referred as the third layer of the mangrove forest. Here, the seawater very rarely submerges the main trunk but is always inundated by water and soft mud. As we made our way inland to where the tide typically reached its upper limit, our navigation and advances became almost effortless. There were minimal obstacles to pass through. I noticed the protruding aerial roots of the *Sonneratia* and *Avicennia* were definitely absent by then.

The ground looked somewhat drier and easier to walk on further inland. Only during high tides do the mangroves here become inundated with seawater. If not the roots or mud, the heat and humidity would soon hit you. I was drenched in sweat and drawn to decide on relaxing amongst the *Rhizophora* trees. They still dominated the second green belt of the mangroves as we advanced inland. Here, another oddity of the mangroves just amazed me. *Most remarkable of all mangrove adaptations must be their fruiting behaviour,* I self-assuredly opined. This after I caught sight of several large slender seedlings of *Rhizophora* dangling from the branches. Their shape and reproductive parts didn't look anything like that commonly seen in the plant kingdom. Each of the seedlings is a brown globular knob with a long green spear protruding from one end. About one foot long and one inch thick, each spear invariable pointed straight to the ground below.

Mangrove plants are viviparous. Unlike most flowering plants, which disperse through dormant resting seeds, mangrove plants produce propagules—seeds that germinate while still on the parent plant and are dispersed via water.

We all have also learned in biology classes that a seed must germinate to give rise to a new plant. This process of germination involves five steps. Firstly, the water fills the seed in what is called an imbibition process. Secondly, the enzymes found in the seed are activated by the imbibed water, which also marks the beginning of plant growth. Thirdly, the seed grows using its food reserves and starts to access water from the ground. This is followed by the fourth step, when the seed forms soft, pliable shoots able to squeeze through the soils and grow towards the sun. The final step occurs above the ground when the pale-coloured shoots turn into deep-green leaves equipped with chlorophyll able to produce food for the plant through a process called photosynthesis.

A point to note is that all the five steps of germination described above occur after the seed is detached and dispersed from the mother plant. This is true for almost all plants except a few. Many amongst mangrove plants are exceptions to this rule. Unlike most plants, whose seeds germinate in soil, many mangrove seeds germinate while still firmly attached to the parent tree. Mangrove plants such as *Avicennia, Bruguiera,* and *Rhizophora* exhibit this rare phenomenon called vivipary, which means their seeds germinate while still attached to the plant. The term *vivipary* means 'live birth' in Latin. It occurs when the young embryo within the seed grows while still on the tree. Once germinated, it develops into a ready-to-grow seedling called a propagule. At this stage, it is already capable of producing its own food via photosynthesis.

Remarkably, the seed is able to break through its protective coat and emerge out of the fruit wall while still attached to the parent plant. There is a good reason for doing so. This ability to reproduce from seeds while still attached to the parent plant is an adaptation so crucial to living in the mangrove environment. The species of *Rhizophoraceae* are well known for having long torpedo-shaped propagules. Essentially, it is already an embryo. Under its own weight, the embryo falls off the tree to start a new plant. During low tide, its pointed end falls off directly, plunging deep into the soft mud, securely anchoring itself ready to start a new life. Not every time it falls off would its pointed hypocotyl end ideally pierce the mud. When it falls but doesn't stand upright, the longish embryo is capable of bending itself to be vertically implanted in the mud. During high tide, it falls in the water initially and floats horizontally to be carried around. The seedling can be water-borne and remain viable over long distances until it finally anchors itself in a suitable substratum. The seedling then would suck the nutrients for growth of the hypocotyl, and this would become the site of a new mangrove away from the parent plant. This by itself is very unlike most plants. The distribution of mangrove species is not through dispersal of seeds but much more advanced seedlings. Viviparous germination is another product of biological evolution that has enabled mangroves to survive. They have successfully overcome the adversities of living in harsh and inhospitable environments, which no other plants can.

The spear-shaped propagules of mangroves are able to float and be dispersed away from the parent plant until they anchor themselves in the mud to start growing and colonising a new area.

From where we rested, I strained my eyes to again form the familiar impression of what a mangrove swamp was all about. It was unlike what most people usually paint pictures of mangroves in their mind. They are far from a dull world of grey and pale green. Indeed, the mud was slimy and blackish in colour, and the malodorous air could be unpleasant. Instead, amidst the combination of punishing humidity and heat, I gasped at the random complexity and resilience of the entire ecosystem. The adaptations of animal species living amongst the half-submersed, time-tested tangled roots and stems are miraculous, almost magical. The sheer brilliance of nature is hard to ignore.

On either side of the mangroves, breaking waves relentlessly cast up soft white sand onto the existing beaches. Familiar species of coastal vegetation, not necessarily associated with mangroves, are found growing along the beaches. They further stabilised the ever-shifting sand banks and beaches in the area. Erosion and degradation of the beaches here are kept to a minimum by the presence of these plant species. Acting like a fortress along the opposite banks of rivers and streams feeding the mangroves were the *Nypa* palms. There were also growth patches of the spiny-leafed *Pandanus*, the needle-leafed *Casuarina*, rapidly spreading *Ipomea*, and the majestic shady *Barringtonia* trees. Fringing the sea, together these coastal plants served as a robust natural barricade against erosion and intrusion of salt water further inland.

A yearning for an enjoyable relaxing holiday in the tropics always conjures up images of lazing in the three elements of nature, sun, sand, and sea. In the panoramic background would invariably be images of graceful curving trunks of the coconut palm *Cocos nucifera*. Arches of their green fronds waving welcomingly to visitors to the tropical islands and beaches have become almost iconic. But there is another plume of arching fronds that create a typical landscape of a tropical escapade. This is the instantly recognisable seashore pandan or screw pine, *Pandanus tectorius*. There are about six hundred species of pandanus spread across the tropics, inhabiting swamps and forests, from the mountains to the seas. It is a small tree with a strange stilt-rooted trunk found growing on beaches and in mangroves. Unlike its close cousin, the aromatic pandan *Pandanus amaryllifolius*, used widely for flavouring in the cuisines of Southeast Asia, the leaves of seashore pandanus lack fragrance; their long narrow drooping leaves have sharp spines along the edges. I found its other common English name, screw pine, rather intriguing. My quick research into the origin of this name revealed that the first word *screw* comes from the tuft of prickly leaves arranged around the trunk akin to the spiral thread of a screw. The second word *pine* comes from the paltry resemblance of its fruit to those of the pineapple. Interesting indeed. Both the pineapple and the pandanus bear compound fruits of comparable size, and when ripe, the pandanus fruit displays bright-orange red almost identical to that of pineapples. Culturally, seashore pandanus leaves, known in Malay as *mengkuang*, feature prominently in the traditional life and handicrafts of the people having access to this natural resource. The pandanus leaves are used in weaving mats, hats, baskets, bags, and ropes. The leaves are allowed to dry slightly over a fire for pliability before being cut into strips. They are left to soak in water for up to four days with a few rounds of replacement before they are ready for bleaching in the sun. Often, chemical dyes are used for a variety of colours. The manual preparation before use can be tedious and time consuming. It is understandable that over time, easily available and cheaper materials for local handicrafts have been sought. Today, plastic strips have widely substituted the traditional use of mengkuang. The immaculate adorable hats, handbags, food coverings, and floor mats woven from pandanus leaves have become almost non-existent.

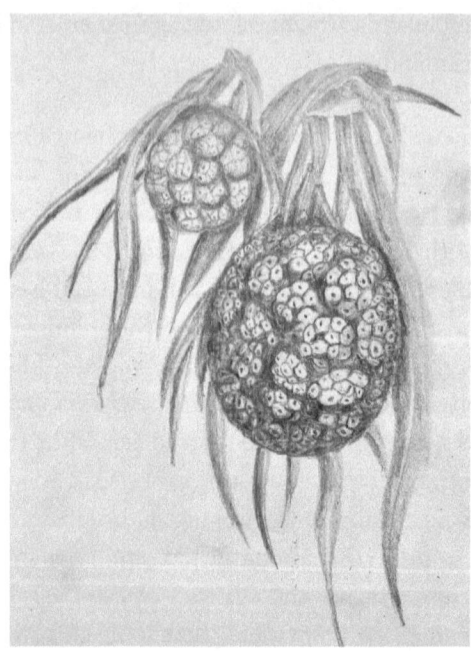

The common name Screw-pine for *Pandanus tectorius* comes from the spiral arrangement of its prickly leaves around the trunk similar to the thread of a screw, and, interestingly, *pine* comes from its fruit resembling pineapple both in size and orange-red when ripe.

In awe, I scanned the long water edge dominated by rows of Nypa palms, *Nypa fruticans*, the solitary palm species that has truly adapted to the harsh mangrove ecosystem. Its attractive female flower comes in the form of bright-red inflorescence consisting of numerous nuts that are somewhat swollen and globular in shape. Despite their thickly fibrous and woody nature, these nuts, when ripe, are able to float. They can be carried for some distance away from the parent plants by the tide. Some even germinate while still waterborne. *Nypa* palm has been known for a long time to have successfully adapted and flourished to the present era. Fossils of *Nypa* pollen have been dated as far back as seventy million years ago.

The young seeds of Nypa palm, *Nypa fruticans*, are edible. The soft and tasty internal part tastes almost like that of coconut flesh, and the sugar-laden sap is often fermented into alcoholic beverages.

Growing up in a village, I was familiar with the *Nypa* palm for its feathery fronds used as roofing materials for thatched houses and farm shelters. My botanic friends found *Nypa fruticans* interesting taxonomically. It is a monotypic taxon—meaning, there is only one species in this *Nypa* genus. But for microbiologists like me, the unusually rich sugar content of *Nypa* sap was potentially exciting. In Southeast Asian countries, palm wine or toddy is an alcoholic beverage produced from the fermented sap of *Nypa* palm since time immemorial. As a child, I remember collecting the palm sap from bamboo tubes placed under the unopened inflorescence of Nypa palm. That used to be a worthwhile effort to treat ourselves to the heavenly sweetness of Nypa sap. But we were informed enough to do this before fermentation took place. The fermentation starts as soon as the sap flows into the bamboo pitcher carried out by the yeast naturally present in the environment. After twenty-four hours of fermentation, the ethanol content can be as high as 9 per cent. My friends and I would have to be quick and timely to taste the sweet sap, because if left for too long, the sap would become acetic acid tasting like vinegar. Today, all I can think of is the potential of fermenting *Nypa* sap into ethanol or butanol, which could provide the long sought-after solution to cheap renewable substrate for biofuel. *A very provocative prospect*, I surmise.

In an ecosystem consistently inundated by water twice a day, some plants could not rely on the soil to obtain a good supply of minerals and nutrients for growth. They evolved other means to nourish themselves including turning carnivorous. A group of plants, the *Nepenthes*, are capable of digesting the soft parts of insects and other animal matter to obtain nutrients for growth. The Malaysian rainforests, especially of Sabah and Sarawak, are often referred to as the epicentre of species diversity for *Nepenthes*. These two states of Malaysia harbour greater diversity of *Nepenthes* species than anywhere else in the world. In Peninsular Malaysia alone, a total of at least eleven species of *Nepenthes* have been documented, and thirty-nine species have been recorded on the whole island of Borneo. This is impressive considering that the total number of *Nepenthes* species worldwide is not greater than 120.

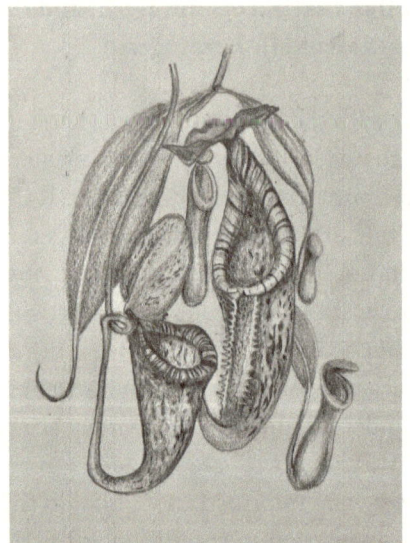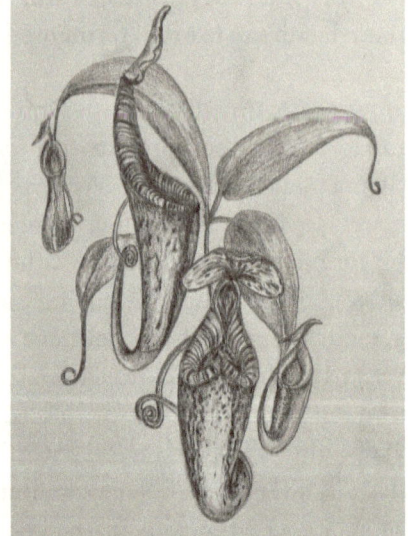

The leaves of carnivorous plant *Nepenthes* are modified into pitchers. Insects and other small invertebrates are lured and trapped in these pitchers containing an array of hydrolytic enzymes capable of digesting the victims for the nutrients required for growth.

A few feet away at eye level, I spotted hanging cups of a *Nepenthes* species commonly found in mangroves. This particular one was *Nepenthes ampullaria*, commonly found in the mangroves and other wetlands. Other species often seen growing in the lowland areas in locations where soil nutrients are poor include *N. mirabilis*, *N. gracilis*, and *N. rafflesiana*. Their pitchers are actually modified leaves. As the plant climbs by the tendrils formed at the tips of their long flat leaves, the pitchers start developing into cup-like structures. The lid remains closed until it opens after a few days to be quickly filled with fluid. A common fallacy is that the lid closes upon its prey. But in actuality, this doesn't happen. The lid has two functions. Firstly, it is a seductive device. Flying insects are attracted to the nectar-secreting glands liberally present on its inner surface. Some pitchers are brilliantly coloured to advertise its nectar, and most *Nepenthes* offer a convenient foothold for the insects to explore further towards the inner surface of the cup. Secondly, the lid forms a rain-protecting canopy over the mouth of the pitcher. This is important to prevent rainwater from diluting the content of the cup, losing its nutritious properties. Immediately below the pitcher opening is a waxy area where nectar is produced. Ants and winged insects on exploratory missions alight on this waxy surface looking for nectar. Inadvertently, they tend to slip and fall into the pitcher because of the precariously smooth and slippery nature of the surface. They lose their grip and fall in the chasm containing acids and potent enzymes secreted by the microscopic glands of the pitcher. The victims drown in the fluid before being acted upon by myriads of digestive enzymes, and the valuable nutrients are absorbed through the pitcher wall. Digestion process in the pitcher is shown to occur very efficiently and rapidly. The acidic nature of the pitcher fluid is required just as in human stomachs. Digestive enzymes known to be present include ribonuclease, lipase, esterase, acid phosphatase, protease, and possibly chitinase. In nature, a medium-sized fly may be completely digested in two days, leaving only the hard chitinous skeleton. Smaller insects like ants are digested within hours.

For most of us who have been fortunate enough to have explored the mangroves, it seems the plant species are superheroes of the plant kingdom. Eons of evolutionary processes and clever adaptations have made them survivors against all odds. It reminds me of a favourite song of mine by Kenny Rogers, 'The Gambler', which says the secret to surviving is knowing what to throw away and knowing what to keep. That is the winning strategy in a poker game. A similar survival strategy has been adopted by life in the mangroves. It is indeed a poker game of survivors for plant species in this inhospitable environment. Over millions of years of evolution and adaptation, every species here has thrown away the inferior features and only kept the superior ones, which can help them survive. The best clones win.

3

+ ✦ ✦ ✦ +

Exclusive Residence

During close to four decades of pursuing my career as a researcher and educator in Borneo, mangroves have been amongst the many fascinating natural ecosystems I grew to admire most. They are definitely amongst the top three I thoroughly enjoyed, in addition to lowland rainforests and coral reefs. Bako National Park is one of the unique places that presents an opportunity to explore the many facets of both terrestrial rainforest as well as mangroves. This protected park, established in 1957, is the pride of Sarawak, a Bornean state of Malaysia, in showcasing biodiversity and ecosystem conservation.

We arrived at the park after an hour's bus drive from Kuching City and a thirty-minute boat ride from Kampung Bako. Situated at the northern part of Muara Tebas peninsula, it covers an area of around twenty-seven square kilometres. Despite its small size, Bako National Park is a gem with respect to conservation of almost every type of vegetation found in Borneo. Besides the famously rich and diverse lowland dipterocarp rainforest, the park boasts of having over twenty-five categories of vegetation from seven distinct ecosystems. These include beach vegetation, cliff vegetation, *kerangas* or heath forest, mangrove forest, mixed dipterocarp forest, *padang* or grasslands vegetation and peat swamp forest, plants, and wildlife galore. The natural landscapes, too, are varied and spectacular. The coastlines feature small bays and coves that stand out as stunning geological formation of limestone outcrops and sandstone cliffs. I found myself standing on a stretch of white sandy beach watching the setting sun slowly disappearing behind the tallest mountain in Sarawak, Mount Santubong. It is no wonder Bako National Park has emerged as a popular destination for both local and international tourists. I came with a specific interest in having a close encounter with the proboscis monkeys known to reside in the mangroves here. Approximately three hundred individuals are estimated to be found inhabiting a small portion of the park. By way of comparison to other places harbouring proboscis monkeys, that is a high-density occurrence, which makes natural encounters with them likely.

Of all monkeys I have the most admiration for, it is the proboscis monkeys. They seem calm and peaceful, unlike the playful, boisterous macaques. Despite their large size of up to twenty kilogrammes, proboscis monkeys are not territorial. They will tolerate other groups in their feeding areas. In this respect, they are often considered one of the most docile species of primates. Sometimes, they are too accommodating that macaques and orangutans have been seen to displace proboscis monkey at feeding sites.

Sexual preferences in humans are a private matter, and usually best kept between you and your partner. For most animals, however, evolution has made their sexual preferences an open secret. A big long nose is considered sexy in the world of proboscis monkeys, giving males with the longest nose a better chance of attracting mating partners.

Its facial features, however, are the least flattering, if you are looking at guises and proportionate appearances. The ridiculously bulbous long nose hangs down the male proboscis monkeys, covering half of their faces. Weird and outlandishly disproportionate, it looks comical. Proboscis monkeys could take the trophy for the ugliest-looking creatures amongst the monkeys and apes of Borneo. They are found nowhere else but in Borneo. Indeed, a rare yet unmistakable denizen of the mangroves. The local tribes of Indonesian Borneo call them *Monyet Belanda*, which means 'Dutchman monkey', as a joke on European traders and colonials of the past era. Both the Dutch men and proboscis monkeys were seen as having lots of body hair, a big red nose, and were usually pot-bellied. The monkey's nose is so long that it may hang down right over the mouth. He has to push it up and out of the way when he eats. And when he looks up to select a leaf, it flops back and smacks him between the eyes. Its scientific name, *Nasalis larvatus*, is derived from this trademark nose. Nasalis is the name of the small muscle on each side of the nose capable of constricting and making the nose flare with gaping nostrils. The male's nose is much bigger and longer, measuring up to almost twenty centimetres, while the females possess a dainty and pointed nose that appears to tilt upwards instead of pendulous.

In the mangroves, proboscis monkeys live in a harem comprising one dominant male and around five or six sexually mature females. Group size varies, with some up to thirty individuals including their offspring. They are highly arboreal, living primarily on leaves that make up 95 per cent of their diet. Because leaves are so poor in nutrients, a proboscis monkey has to eat a large amount of leaves. Hence, it has to spend most of its time looking for food and eating until its stomach contents can make up one-quarter of its body weight.

This excessive eating is not habitual greed or gluttony at all. It is in fact a unique diet adaptation that enables the species to live in the mangroves. Most animals find the mangrove leaves unpalatable or even mildly poisonous. But for the proboscis monkey, the mangrove leaves are just another meal to savour. Its pot belly holds several stomachs, each filled with highly specialised bacteria capable of digesting the tough cellulosic content of mangrove leaves. An adult can munch in excess of 1,500 of the bitter mangrove leaves known to contain high level of tannins. Like cows and other foliage eaters, the large stomach of the proboscis monkey is divided into compartments equipped with fermentation chambers where digestion takes place. Here, the breakdown is facilitated by special cellulose-digesting bacteria. Their stomach is twice as large as other colobines with similar digestive systems. This gives them a permanently pregnant look, even the males. They spend most of their time jumping from tree to tree feeding. Only very rarely would they go to the ground or riverbanks when they needed to drink. Otherwise, they would get all the water needed from the leafy diet. Mother Nature must have worked her way to win the heart of the proboscis monkey through his stomach. Why not? It has proven to be a useful adaptation to reduce competition for food from other animals living in the vicinities of the mangroves.

Proboscis monkeys are endowed with complex multi-chambered stomachs, allowing them to survive mainly on a diet of mangrove leaves, seeds, and unripe fruit. For digestion of this fibrous cellulose-rich food, they rely on a host of bacteria living symbiotically in their stomachs.

The plan for the second day at Bako National Park was to comb the intertidal zone and survey the diversity of life inhabiting different microhabitats there. The sun had climbed to almost directly overhead. The morning tide was fast ebbing. I could almost visualise the speed of waterline receding towards the horizon. After breakfast, the rest of the team took it easy, making their own way to seek out different microhabitats of the mangroves. I had the entire mid-morning hours to kill. So I decided to start exploring the long stretch of shady mudflats close to the estuary until lunchtime.

Carefully placing my bare feet on patches of drying mud, I moved aimlessly towards the east side of the mangroves, bending over and squatting to pick up broken shells of bivalves, scrutinising floating debris of mangrove fruit and flowers. I stopped to rinse my sandy fingers in puddles of seawater on the beach and indifferently picked up a twig and poked into holes I saw in the sand. I was half-expecting some frightened creatures might just scuttle up to the surface and challenge me, the intruder. On the smooth sandy bank about six feet in front of me, some little creatures just leapt forward; quite an impressive distance for something only about five or six inches long. It was a party of mudskippers, a kind of fish that seemed happy to spend more time on land than in water. They certainly have acquired novel means to breathe, move, and see better on land. But like conventional fish, they could also remain submerged in water indefinitely. Amazing adaptation. They owed this to a pair of enlarged gill chambers, which enabled them to store a large volume of water and air when on land. Their gills were kept forever moist, a prerequisite enabling them to filter the water-air mixtures efficiently. This way, they were assured of a regular supply of oxygen to all their organs when out of the water. Although having the typical appearance of any other fish, the frontal pair of fins allow the mudskipper to skip across muddy surfaces and even give them the ability to climb trees, aerial roots, and low branches.

The exposed mudflats were buzzing with activity. I saw silhouettes of creepy crawly things of different shapes and sizes scurrying busily against the shimmering background of the soft sand. Some almost magically disappeared into the sand, and some chose to hide inside holes temporarily. Yet some languidly moved around as if tasting and feeding on make-believe food along their paths. I hastened to investigate. They appeared to haphazardly scurry away sideways instead of forwards or backwards. Instantly, I recognised dozens of fiddler crabs. They seemed to be foraging on something invisible in the sand; probably having a feast on micro-faunal organisms including marine bacteria, algae, and fungi. All appeared to be engaged in scraping off and feeding on something edible from the sand particles. The primary diets of fiddler crabs are detritus from decaying plant and animal matter in the environment.

A male fiddler crab performs bizarre gestures to catch the attention and interest of a mate. Standing at the entrance of his freshly dug home, the soliciting male would energetically hold his heavy oversized claw in the air waving back and forth. A successful seductive performance would end up with the mate drawn into his burrow for mating.

Remembering David Attenborough's documentary film about the somewhat bizarre mating ritual of fiddler crabs, I elected to sit and watch. The ritual was slowly unfolding under my intense scrutiny. The male fiddlers were seen forming a line alongside their freshly dug burrows and doggedly waving their large claws back and forth akin to a celebrity violinist in a trance. No wonder it is named a fiddler. I found it delightful to watch the male crab performing this waving of his gargantuan, oversized claw with aplomb and purpose. I was reminded of a painting *The Violinist at the Window* by Henri Matisse. The fiddling motion was meant to seduce females and draw the female's attention. The soliciting males would also tap the ground with their claws for added attention. Success in this seduction process would end with the receptive females drawn inside the burrow and mating ensued. She would only emerge two weeks later to release her fertilised eggs into the water. *Bizarre or not, all nature's creatures join together and mate to express nature's purpose—that is, for the continuation of own species*, I philosophised as I shook my head in wonderment.

I could feel the intense heat of the midday sun on my neck. Without thinking, I found myself under a shady area of the mangroves along a scrubby bank. I noticed some burrows along some parts of the bank that had become exposed at low tide. I was almost certain they belonged to the mud crab *Scylla serrata*, a highly exploited species found in sheltered waters of estuaries and mangroves. Like other crabs, this species of mud crab has a complex life cycle, which includes larvae, juvenile, and adult phases. Adult crabs come out to feed in shallow areas below the low tide mark at night. Most of the day, they bury themselves in the mud, staying clear of the hot sweltering sun. Here, their diet comprises mainly shellfish and crustacean species. Their growth rate is slow, around ten centimetres in width annually to become sexually mature in about two years. The typical life span of a mud crab is thought to be three to four years. Today, mud crabs are in danger of over-exploitation. They are amongst the most commercially valuable crab species in the world. The bulk of their market value comes from selling them as live seafood for consumption as delicacies in restaurants.

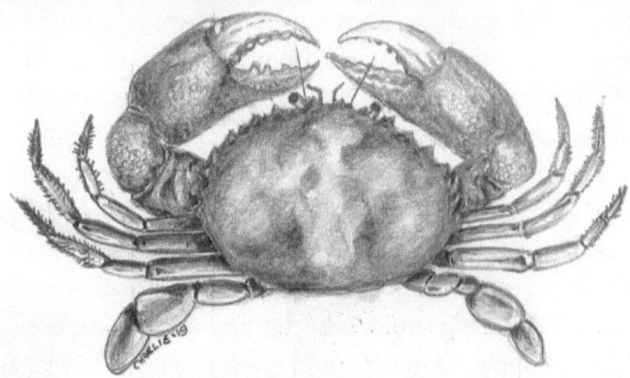

Mud crabs displace sediments and mix particles when they create their burrows. Ecologically, they are known for this role in engineering the mangrove's physical and chemical properties. They help to increase the surface area of mangrove sediments, providing more oxygen for bacterial communities to break down organic material and recycle nutrients.

Farming of mud crabs for consumption remains the only hope of avoiding wiping out the populations of this species from their rapidly deteriorating mangrove habitats. The basic biology for aquaculture production of mud crab was not fully understood until a few years ago. Information on their growth requirement, hatchery, and nursery technology was scarce, requiring more research. I recalled a research project undertaken by a former student at my university in Sarawak a few years ago. He encountered massive problems trying to culture mud crab, *Scylla serrata*, because of cannibalism. The adults were feasting on smaller individuals during growth. Much to the student Ikhwanuddin's surprise, cannibalism was the greatest cause of mortality in his attempt to grow mud crabs before reaching marketable size. This savage act on the smallest size juveniles was extremely high in the presence of the larger adult crabs. This could be prevented by manually and tediously separating them in cages when they reached certain sizes. Today, Professor Ikhwanuddin is a regional expert in mud crab culture regionally if not internationally.

I have been totally immersing myself within the magical elements of the mangroves, busily hobnobbing with nature. It gives me great pleasure to be conscious of the fact I was surrounded with plants and animals that have been the products of evolution and adaptation over millions of years. To understand and appreciate that alone made me more appreciative and passionate about nature and the environment I live in. Once I was with a group of students in the mangroves off a small coastal settlement, Lok Kawi, in Sabah. Excitement and shrills filled the air as one of the lecturers from Universiti Kebangsaan Malaysia Sabah (UKMS) turned up with a beautiful slender black-and-yellow striped snake in hand. It was the yellow-ringed cat snake, *Boiga dendrophila*. The species, commonly referred to as the mangrove snake, is endemic to Southeast Asia including Cambodia, Indonesia, Malaysia, Myanmar, the Philippines, Singapore, Thailand, and Vietnam. It averages around two metres in length. But the one calmly held on exhibit in a tight grip at the neck was above average, close to three metres in length. The biggest and longest *Boiga dendrophila* I have seen was at a weekend open market in Sanam Luang, Bangkok, Thailand, which was around four metres long. I didn't ask if the sale was for a pet or food to be consumed.

Seeing the gorgeous glistening reptile slithering and struggling to be free, the spontaneous question asked by students at the time was 'Is that poisonous?' A cool, calm, and collected reply, 'Yes, of course', was from the snake charmer Robert Stuebing, an American lecturer who came to Malaysia as a Peace Corps volunteer in the early 1970s but stayed on to teach zoology at various local universities. 'It's venomous all right. But it won't kill a man,' Rob quickly explained to the wide-eyed students. 'They are only mildly poisonous. Their venom could, however, result in intense swelling, but no fatality has been reported from bites of mangrove *Boigas*.' That assuring qualification by the expert managed to draw the students' gaping mouths shut immediately.

The venom of yellow-ringed cat snake, *Boiga dendrophila*, has not been known to kill a man, but its bite can result in intense swelling and discolouration of the skin. This mildly poisonous rare-fanged snake is designed more for killing of amphibians and reptiles.

Saltwater crocodiles, egrets, herons, flying foxes, and monkeys are all exclusive residents of the tropical mangroves, just to name a few. Each of these species could be encountered when they come out in full activity and commotion during the day or at night. The mangrove fireflies, however, can only be seen at night. I was amongst the visitors converging on Kuala Selangor to watch the spectacular synchronised flashing of thousands of fireflies along the banks of Selangor River. It was only a couple of hours' drive from the capital, Kuala Lumpur. Every evening, the firefly belonging to the beetle family, predominantly *Pteroptyx tener*, would congregate on the mangrove trees *Sonneratia caseolaris*, known as *berembang* in Malay, putting on a fairyland display of lights. Because of its close proximity to and ease of access from Malaysia's capital city, the Kuala Selangor mangrove has become the go-to destination for international nature lovers. A wonderful experience for those seeking to witness and experience a natural phenomenon uniquely found along the water's edge of tropical mangroves.

After a hearty meal of rice, fish, and stir-fried mixed vegetables, we ventured out to marvel at this awesome firefly light show we have read so much about. The brilliance on the distant horizon seemed to dim earlier than the previous day. I could hear chatter, not of insects or birds but of team members exchanging notes of the day's encounters and observations in the field. The atmosphere in the boat was abuzz with exchange of information on fireflies. Similarly, these are the facts. There are more than two thousand species of firefly found in temperate and tropical environments around the world. Light production in fireflies is due to a chemical reaction that occurs in specialised light-emitting organs located on the lower abdomen. The enzyme luciferase acts on luciferin in this organ to stimulate light emission called bioluminescence. Its function in the adult firefly is primarily to locate other individuals of the same species for reproduction.

Our boatman politely gestured for everyone in the boat to turn off their headlights. The boat silently floated along the riverbank in dimness as the clarity of vision was made more intense in spotting the twinkling lights of the celebrity insects. It was pitch darkness initially, but gradually my eyes adjusted well enough to make out the silhouettes of trees lining the riverbanks. I gasped at the discovery that, under darkness, the mangroves came alive with an array of luminous creatures. The wooden paddle was gently lunged into the water; with just enough force to glide our boat forward. Phosphorescent planktons briefly showed up even from such watchful disturbance of the water surface. Everyone in the boat was suddenly gasping in awe at the sight of a swarm of fluorescent green jellyfish gently whipping their way in the water not too far from the rear side of our boat. Small fish darted away from the wake of the boat. I could almost touch the flutter of bats hunting mosquitoes and moths above my head. A chorus of whines and calls of frogs, cicadas, and crickets sounded a couple of notes higher in the stillness of the night. Across the bank, there was a gentle rustling of the mangrove leaves as we approached. *Nocturnal creatures, crabs, rats, or otters could be out rummaging and feasting*; my mind threw some wild guesses. Fluorescent mushrooms glowed from amongst the driftwood on the bank slopes in the hope of being spotted, eaten, and getting their spores dispersed by ground rodents.

As in other bioluminescent species, jellyfish and mushrooms can glow in complete darkness. They contain a unique protein luciferin, which glows when exposed to oxygen in a chemical reaction catalysed by the luciferase enzyme. Interestingly, jellyfish glow to ward off predators, while mushrooms glow to encourage foraging by animals, which can assist in the dispersal of its spores.

I caught sight of small yellow lights flickering and floating away into the darkness. They were flashing in synch with one another. These were fireflies of the genus *Pyrophorus*. They were soft-bodied beetles of five to twenty-five millimetres in length. On the underside of the abdomen, fireflies are equipped with special light organs capable of glowing in luminous flashes. The females produce the brightest light. The flicker of light is a means of communication—a sort of insect Morse code. Each species employs a specific signal: the flashes produced by one female will attract the male of the same species only. As darkness grew, the message from the few initial 'Morse codes' appeared to be receiving responses from all around. Branch to branch, one *Sonneratia* tree after another, the luminosity multiplied. The flickers appeared haphazard and uncoordinated at first, but progressively became synchronised. In a matter of minutes, the entire row of trees lining the riverbank was pulsating on and off. It was simply magical. A sight to die for!

On the underside of the firefly abdomen, there is a special light organ capable of glowing in luminous flashes.

Along the banks of the river feeding the mangroves, a spectacular celestial display of sparkling lights has been turned on each night in the past millions of years.

I found such dreamlike display by Mother Nature immensely provocative and moving—a multitude of species bringing about a repertoire of delightful natural phenomena. Every trip to the mangroves has helped me to understand the uniqueness of the mangrove ecosystem and what is to be found there. Undoubtedly, the onus is on us to act responsibly to sustain their continued existence.

4

Visitors Welcomed

In a biological ecosystem like the mangroves, some species are restricted to their own niche to feed and reproduce. A niche is a specific area the animal occupies in an ecosystem. This living space is defined by a range of resources and conditions that allow the species to maintain a viable population within that space. We can view a niche as the part of ecological domicile where both the biological and physical factors within help maintain the survival of a particular species. Several species may coexist in a niche. Animals of different species can interact between themselves positively to make the space conducive for everyone's survival. Together they function to maintain the niche, which can meet their respective needs and survival advantages. They can interact with respect to accessibility to sufficient food resources, reducing competition or avoiding predators.

The niches found in mangrove forests between land and sea have led to some unique advantages for the continued survival of a number of species. Many have made these niches their lifetime homes as they evolve and adapt to the environment over generations. They eventually become exclusive residents found only in the mangrove forests and nowhere else. Our familiar proboscis moneys, mudskippers, fiddler crabs, and mud crabs discussed earlier are some of these elite residents. In some mangrove forests I have explored, there are other animal species that can be regarded as visitors to the mangroves. These are species that have not evolved and adapted specifically to withstand and survive the harsh conditions of the mangroves in totality. They can survive just as well elsewhere. But they are smart enough to recognise that mangroves harbour plant and animal species they can forage for food. Thus, they can effectively reduce competition from other inland species in their usual habitats.

At Bako National Park, Sarawak, the silvery langurs, *Trachypithecus cristatus*, are such opportunistic residents. Also known as the silvered leaf monkeys, they are often found in mangroves and coastal and riverine forests. They are specialist folivores, which means a major portion of their diet is leaves. Fruit, seeds, and flowers are also on their menu, but 80 per cent of their diet is leaves. In the rainforests, the silvery langurs are known to feed on tough mature leaves considered unsuitable for consumption by other species of monkeys. Like other langurs, silvered leaf monkeys possess dedicated digestive systems that allow for the fermentation and break down of leaves akin to the elite mangrove residents, proboscis monkeys. For silvery langurs, the tough leaves of the mangroves would be just another leafy meal. They are equipped with a chambered gut able to handle the tough cellulose fibres and high-tannin content of mangrove trees. By moving their group to feed in the mangroves, they avoid competition for food with other langurs found inhabiting the terrestrial rainforests. They usually travel in search of food in groups of ten to forty individuals. Their social structure is definitely that of male dominance, much like a harem of the proboscis monkey. The adult male overbearingly takes charge, checking the safety and affairs of several females. Infants also travel with the group to be cared for by the females.

In general, primates are mobile because their body form allows different adaptations for moving around. During foraging, they are able to navigate and maintain both arboreal and terrestrial lifestyles. A set of anatomical and behavioural adaptations become crucial, including binocular vision, elongated limbs with claws for gripping and swinging, and a tail for counterbalancing.

Outwardly, an adult silvery langur possesses a combination of a fearsome as well as a comical appearance. The face is covered with lush silvery grey hairs, which seem to end as a pointed crest on top of the head. From within these dark-grey furry faces, a pair of huge bloodshot eyes stare out at you. They aren't boisterous and unruly at all. Unlike the macaques, which can be intimidating and nerve-wracking in the presence of humans, silvered leaf monkeys are more timid and docile. They seem to prefer a surrounding atmosphere that is friendly and peaceful. In the vast rainforests, the population distribution of silvery langurs is wider than other monkeys. They spread out to different areas, almost deliberately staying quite apart from other monkeys. Even if they are found within the same forest areas, the silvery langurs are found feeding in the middle canopy of the forests, leaving other species confined to the higher branches. While they quietly feed on the mature leaves below, other frugivorous species can enjoy the abundance of fruit found on the canopy above. They aren't intimidated by human presence. Often found mixing with humans, they will watch us expressionlessly, just as questioning as we are watching them. At the same time, they will be nonchalantly busy grooming and playing with other members of the group. But monkeys will always be monkeys. Some mischievous ones have been known to swiftly make a snatch of any item they fancied from humans, especially anything that seemed edible.

At the crack of dawn, I was awake before the final rays of the rising sun. The air was crisp and fresh. I craved for the reigning peace of the morning and chose to be alone before going for breakfast. I swung my camera bag onto my shoulder and wandered outside amidst the glistening dewdrops. I headed towards the limestone cliffs, at the edge of the freshwater catchment area on the east side of the mangroves. Patches of secondary forest growth were evident. Lush foliage of riverine plants was flowering, dotting the usually green monotonous landscapes with rich colours of yellow and orange. The stillness of the morning was most appealing. I heard a ruckus in the branches some ten metres ahead. It was a group of long-tailed macaques, *Macaca fascicularis*, a common species of monkeys considered very successful in adapting to various ecosystems, including the mangroves. Their success as a species can be attributed to their omnivorous diet. Here in the mangroves, they are an opportunistic omnivore, feeding on a variety of animals and plants. They are able to break the rules and adopt a much more wide-ranging diet than the carnivores and herbivores. This way, the species has an added advantage as visitors to the mangroves.

Long-tailed macaques are also known as crab-eating macaques. Despite the name, the crab-eating macaques typically do not consume crabs as their main food source. During visits to the mangroves, they do not just focus on plants but will avail themselves of an opportunity to complement their diet with marine food items. Here, they are always surrounded by an ample supply of crabs and shellfish. During low tide, the macaques venture down to the reefs and wade in shallow waters to hunt for crabs and several other marine life. Clearly having a generalist diet, feeding not only on fruit, nuts, leaves, and insects but also on marine animal species, is another survival gain for the crab-eating macaques.

Long-tailed macaques are omnivorous, feeding on not only foliage and fruit but also animals and insects. They are often found roaming on the beaches adjacent to mangroves hunting for crabs in the intertidal pools, hence also known as crab-eating macaques.

The mangroves along the northern stretch of the park appeared less disturbed, lush, and more fertile in their growth. It was high tide that afternoon. Our boatman steered clear of the shallow and muddy patches and managed to get deeper into the mangroves. The penetration was made possible through a small stream, which appeared to have come from within the massive tangles of *Rhizophora* roots. On a distant tree about twenty feet high, I thought I was looking at clusters of some peculiar black fruit or shrivelled leaves. But I could also hear busy shrills and squeaky chatter coming from the same tree.

Curious, I peered through my binoculars and was greeted by a bewildering sight—roosting flying foxes hanging head downwards in huge numbers. The flying foxes, *Pteropus hypomelanus*, the largest of all bat species, are always seen hanging upside down from branches of mangrove vegetation or palm trees. They are social creatures often seen engaging in licking, cleaning, and grooming each other to express affection. Gregarious nocturnal bats, they feed on all varieties of fruit. Their long bristly tongue is useful not just for licking and grooming each other but also for sucking sweet nectar from flowers and fruit. Unlike their smaller bat cousins, flying foxes roost outside in the sun rather than in caves. Their huge pair of eyes afford them excellent eyesight, freeing them from the need to rely on echolocation in search of food at night. An encounter with hundreds or thousands of roosting flying foxes each having their wings tightly gathered around their heads and torsos can indeed prove an unforgettable experience.

Although often referred to as fruit bats, the flying foxes have voracious appetites for flowers, nectar, and occasionally insects. They are called flying foxes because their heads resemble those of foxes. Unlike other bat species, which can echolocate, they locate food with their keen sense of smell and navigate with keen eyesight.

Another frequent visitor to the mangroves is the Asian short-clawed otter, *Aonyx cinerea*. It is the smallest otter species in the world found only in South and Southeast Asia. Of thirteen species of otters worldwide, this lovable creature weighs approximately only three kilograms. It is no wonder that Asian short-clawed otters are habitual visitors to the mangroves because, like other otter species, it is a semi-aquatic mammal comfortable in both terrestrial and aquatic habitats. They are very social, living in family groups of up to twelve individuals. A highly intelligent animal, they have adapted brilliantly living in shallow streams and water bodies. As carnivores, they thrive on a range of animal species including molluscs, fish, frogs, crabs, and other crustaceans that can be found in abundance in the mangroves. I have watched these highly intelligent animals foraging in the shallow waters of the mangroves on many occasions. It has extraordinarily small claws on its fingers, hence aptly named short-clawed otter. They don't need long sharp claws like other otter species because these short claws serve them well as fingers. They chase after, catch, and grasp their prey very efficiently because of these sensitive and dextrous short claws. Because they are endowed with opposable thumbs, they are able to feed themselves entirely with the front paws. Clearly, the species has evolved the perfect grasp to hold on to the slithering and slippery fish catch most securely before biting off the heads. Because they do not possess the long claws like most otters do, Asian short-clawed otters exhibit less webbing between their digits.

Asian short-clawed otters, *Aonyx cinerea*, are energetic, playful, and constantly on the move while emitting adorable piercing shrills. They are increasingly popular as novelty pets, particularly in Japan. Illegal trade of the species must be strictly enforced to stop poaching in their natural habitats.

Ever been to the zoo and wished the animals there could be energetic and move more? The slow lorises, reticulated pythons, lions, or tigers just seem to lie still all day long, spending their time sleeping or resting in the shade. But not Asian short-clawed otters. They seem to be constantly moving and being active. Amidst all these constant movements, they also emit piercing shrills and seem extremely playful. The entire family is forever moving as a group on land or swimming in the water.

There was some movement at the fringe of the riverine vegetation about twenty metres ahead. I caught sight of a fat Asian water monitor lizard, *Varanus salvator*, about two metres in length. It is one of the most common visitors to the mangroves. Highly adaptable and resilient, this dinosaur-looking creature can survive almost anywhere. At ease dwelling within the mangrove environment, it adapts to roaming around under the shady undergrowth of the mangroves, getting scruffy and dirty in mud at low tide, and cooling off swimming gracefully in the waters at high tide. Having an opportunistic lifestyle, they feed on almost anything and are not particularly choosy in their choice of habitat. Such versatility in foraging and habitation requirements are factors making monitor lizards quite ubiquitous and are found living almost side by side to humans on the outskirts of our cities.

Monitor lizards have been shown to reproduce without a mate. Such 'virgin births' through asexual reproduction, called parthenogenesis, enables monitor lizards to establish new colonies without mating. This might occur when females find themselves washed up alone on isolated offshore islands after a storm at sea.

Lizard species generally have a low metabolism. But monitor lizard's metabolism is known to be high. There is a plausible reason for this. While other species are often seen sunbathing to warm their body temperature or waiting patiently for insects to pass by, monitor lizards are always on the move. They don't wait for food but actively hunt for it. For their size, monitor lizards run fast, powered by highly developed leg muscles. In fact, they can move faster than most of us can run! They are built for speed. Equipped with a powerful body and strong legs, they are capable of running on their hind legs at a high speed, clocking twenty-eight miles or forty-five kilometres per hour. To escape from predators or to avoid humans, a monitor lizard can swiftly dash into bushes, burrows, or even dive under water to hide. Being carnivorous, monitor lizards chase after their prey. They are armed with small sharp teeth, which are used to tear the flesh of their prey. Their claws are equally piercing. Strangely enough, their sharp teeth and claws are not always used to pull and tear their food. More often, they swallow their prey whole. Despite acting in such a vociferous manner, they do not choke on their food. This is because the epiglottis prevents large chunky food from entering the larynx, allowing it to breathe normally while swallowing food.

Skinks, lizards, and other cold-blooded creatures bask in the sun to keep warm, a necessity called thermoregulation. New studies reveal they do it for health reasons also—that is, to regulate their vitamin D. All vertebrates need this fat-soluble vitamin, which has long been known to help the body absorb and retain calcium and phosphorus, both of which are critical for building strong bones.

Generally, monitor lizards rely on their earthy-coloured camouflage to quietly sneak up to and spring an ambush on their prey. They are also able to scuttle away very swiftly when they need to escape from the predators. When forced to a defensive mode, they often resort to viciously use the most intimidating weapon at their disposal—their muscular tail with which they stupefy their prey. Typically, the tail of a monitor lizard accounts for half of its body length. They can lash out in full force their incredibly powerful muscular tail on any invading predators. In water, the tail propels the animal gracefully and silently. On land, the tails are so powerful that they can thrust their entire body suddenly and speedily in pursuit of prey or escaping from predators. During a fight between individuals, monitor lizards hoist up almost the entire length of their body in an upright posture supported by their strong tail. Unlike other lizards, monitor lizards cannot regenerate their missing tails. Only in desperation or pushed into a corner would a monitor lizard use its sharp claws to escape by climbing trees.

A monitor lizard's tail has another use as well. The tail is dragged along the path on the ground to mark its territory, much for the same purpose a dog urinates to mark its territory. These territorial markings serve to protect its hunting ground, food source, cool shelter on a hot day, or a basking area on a cold wet day. Most species of monitor lizards remain on the ground amongst the undergrowth, but others prefer to venture into holes in trees or riverbanks.

A tripod stance is a behaviour in which four-legged animals rear up on their hind legs and use their tails to support this position. Monitor lizards use this behaviour during fighting for food, mate, or territory. Both duelling animals stand on their hind legs and tails, trying to wrestle and overpower each other.

I cautiously approached and followed a monitor lizard's advances on a muddy bank. After a brief slow motion akin to a crawl of a baby, it suddenly stopped and tilted its head up as if trying to make sense of the clicking noises coming from my camera. Suddenly, it scuttled and ran down the water's edge and plunged into the water without making much of a splash. Highly mobile, it swam gracefully, keeping its limbs to the side of its body, propelling itself with its flattened tail. Upon reaching the opposite bank, it quickly disappeared underneath the thick undergrowth.

On several occasions, I would accompany my university professors on their biodiversity expeditions at night. Besides the usual streams and riverbanks of the rainforests, the mangroves could also yield some interesting species of herpetofauna specimens for research and teaching. Spotting frogs, snakes, and lizards in the trees and muddy banks of the mangroves can be more challenging than the terrestrial rainforests as we had to traverse through the tangling mangrove roots. A night walk in the mangroves was also ideal for wildlife sightings of nocturnal species foraging and hunting in the stillness of the dark surroundings. Because the eyes of herpetofauna strongly reflect light, spotting them from a distance was relatively unproblematic. Using lights to scan the water for their eyeshine was the best way to spot them at night. Snakes, lizards, frogs, and crabs were out then and could easily be spotted even by the untrained eyes of first-timers. Sometimes, all that was visible to us would be their gleaming eyes sitting quietly amongst the foliage, mangrove roots, or muddy banks. I have on many occasions seen rarely encountered species of frogs and lizards just sitting like statues on floating logs or forest debris as they slowly drifted past. Their glowing jewel-like eyes were flickering red and blue in the wavering beams of our light. Their motionless, enigmatic gaze and mesmerising eyes appeared weirdly magical—almost comical.

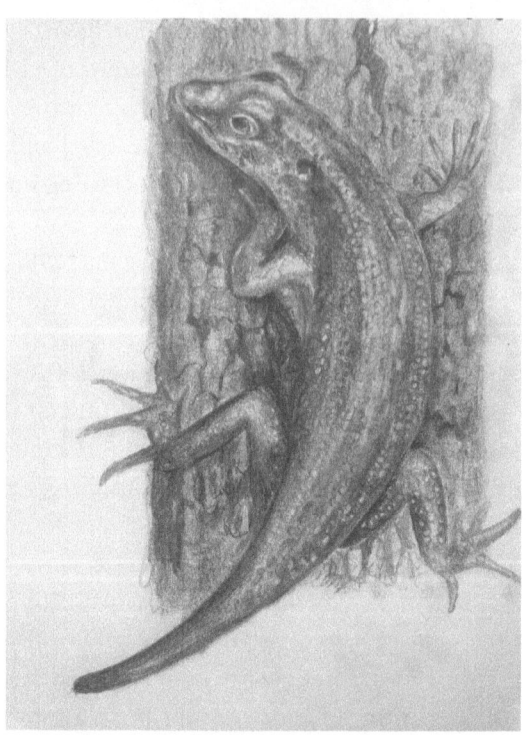

Littoral skink, *Emoia astrocostata*, has an uncommon lifestyle to other skinks in the world. Besides insects on the menu like other lizards, littoral skinks are fond of feeding on crustaceans found in the intertidal zone of the mangroves. They are often seen perching patiently on driftwood or boulders in intertidal pools, waiting to catch unsuspecting victims.

Amphibians are well known as osmotically sensitive organisms because of their highly permeable skin and eggs. They can easily succumb to dehydration in aquatic environments of high salt concentration. Biologists readily dismiss the possibility of seeing frogs in the coastal saline or brackish water environments. But there are exceptions to this rule. The frog *Fejervarya cancrivora* is found inhabiting comfortably in the mangrove swamps of Southeast Asia. This is an extraordinary feat. It is one of only 144 modern amphibians known to tolerate marine environments. The species can withstand immersion in seawater for brief periods and much longer in brackish water. This is achieved by increasing production of and retention of urea in their bodies. In full-strength seawater, the excretion of urea is greatly diminished, causing the osmotic concentration in the blood slightly hypertonic to saline environment This way, the frog as a whole remains slightly hyperosmotic, enabling them to survive in salt water with salinity as high as 2.8 per cent. Therefore, the frog *Fejervarya cancrivora* is able to live and breed in a variety of saline coastal and inland habitats. Remarkably, their resilience in reproducing under such conditions is attributed also to their tadpoles being able to withstand salinities as high as 3.9 per cent. Unlike frogs found near freshwater habitats or banks of inland rivers, which thrive on primarily terrestrial insect species, mangrove frogs living near brackish water predominantly feed on marine animal species. Food for this phenomenal frog include small crustaceans like crabs, hence known also as crab-eating frogs. But *Fejervarya cancrivora* is also locally hunted for food in Southeast Asia, particularly Indonesia and the Philippines. In recent years, farming of this species for their reasonably sized edible legs has become a popular commercial venture in Java, Indonesia.

Fejervarya cancrivora is an inhabitant of the mangroves because this frog can tolerate marine environments and is able to remain immersed in brackish water for extended periods. Adults can survive in salt water with salinity as high as 2.8 per cent, and their tadpoles in salinities as high as 3.9 per cent, which is higher than that of seawater at 3.5 per cent. In Southeast Asia including Java, Indonesia, this crab-eating frog is hunted for its meaty edible legs.

In mangroves that are fed by large rivers, the local population will always remind me to take extra precaution of the saltwater crocodiles *Crocodylus porosus*, the largest reptile known to live in this environment. They can grow to a length of six to seven metres. I usually heeded the warning of local residents and enquired if it was the breeding season for these fearsome creatures as that is when they are naturally more aggressive. The people living in the vicinity of mangroves would normally be knowledgeable as to when courtship began during the wetter months of the year. This was followed by laying their eggs around the second quarter of the year, and ninety days later, the hatchlings would be born. The third quarter of the year would be the time when female saltwater crocodiles are most aggressively protecting their young. This is also the period when human-crocodile encounters are known to increase in Sabah. We were discouraged from disturbing them by steering clear of vegetative locations where their nests were likely to be found.

Saltwater crocodiles are intricately linked to other animals and plants in a web of life. Humans need to coexist with them. Crocodiles should not suffer from the rapid loss of their natural habitat and scarcity of food because of human actions. Ecologically, they play a role in the intricate food web and overall health of the mangrove ecosystem.

In brackish water environments, we know that saltwater crocodiles are intricately linked to other animals and plants in an intricate web of life. This is especially true in mangroves influenced by tidal rivers and streams near estuaries where their food sources are plentiful. They feed on crustaceans, fish, turtles, snakes, lizards, birds, and bats. It is not uncommon to see them going after terrestrial wildlife such as macaques, proboscis monkeys, and wild pigs. Occasionally, we hear of huge crocodiles catching domestic dogs, cows, and water buffaloes that have ventured too close to the water's edge. Despite their huge size, crocodiles are able to lunge forward and move fast on land using their powerful tails. Human attacks are rare but do occur, resulting from a new proximity of human settlements to crocodiles. This is often the result of mangrove degradation, which can no longer support the rich diversity of life, resulting in the scarcity of food for this monstrous crocodilian species. The continued environmental degradation caused by human inhabitants remains a threat to crocodile habitats. Pollution and siltation of rivers further diminishes food availability and further restricts the movement of crocodiles. Under such adverse scenarios, the risk of crocodile attacks will continue to heighten. Indeed, we need to coexist. Saltwater crocodiles should not suffer from the rapid loss of their natural habitats and scarcity of food as they play a role in the intricate food web and general health of the mangrove ecosystem.

In 1983, I participated in a workshop on marine algae organised by Commonwealth Scientific Council in Cairns, Australia. I took the opportunity to attend because of the rare chance for me to experience scuba diving at the top diving destination in the world, the Great Barrier Reef. The workshop was carried out primarily at Orpheus Island, located just off the coast of North Queensland where most parts are pristine national parks. I have dived in most coral reefs in Sabah, Sarawak, and Peninsular Malaysia but was always told of the magnificence of this World Heritage site, Great Barrier Reef. I was indeed humbled by the size, variety, and colours of the soft corals, sea fans, and gorgonians I saw there. I found the reefs were more radiant, erupting hues and colours than I have seen in Malaysian waters. At night, the participants were taken to explore the mangroves on the island. Vivid in my head, I can still picture a phenomenon of nature I had not seen in my years of combing the shores of the tropics. An amazing spectacle was an en masse arrival of horseshoe crabs to lay their eggs on the beach. Horseshoe crabs are a living fossil—a survivor of evolution. The ancient creature has remained unchanged in appearance for 350 million years. It is not a true crab but more closely related to spiders and scorpions. The name is derived from its gross appearance resembling the shape of a horseshoe. As formidable and grotesque as the creature may look, a horseshoe crab is not venomous; neither has it a sharp weapon it may use. That particular night on Orpheus Island happened to be one of the rare peaks of a spawning season for horseshoe crabs. Thousands of them arrived on the sandy beach—a bizarre phenomenon of Mother Nature that has remained etched in my memory till today. Such spawning events usually take place during the high tides of full and new moons when the water level is highest on the beach. The beach of Orpheus Island was adequately protected from the harsh wave actions, making its sand and pebble mixture perfect for the incubation of horseshoe crab eggs. As far as my eyes could see, female horseshoe crabs crawled up to the high waterline on the beach with a male attached to each of them. In addition to the piggybacked male, several other males were also seen on and around the spawning couple, attempting to fertilise the female's eggs. A female may have five or more males attempting to mate with her in a single egg-laying episode. In an evening, a female crab can lay several egg clusters. Throughout the spawning event, she may spawn repeatedly over several nights to lay one thousand or more eggs. After hatching, horseshoe crabs spend their first few years of life on the tidal flats and move out farther from the shore as they age.

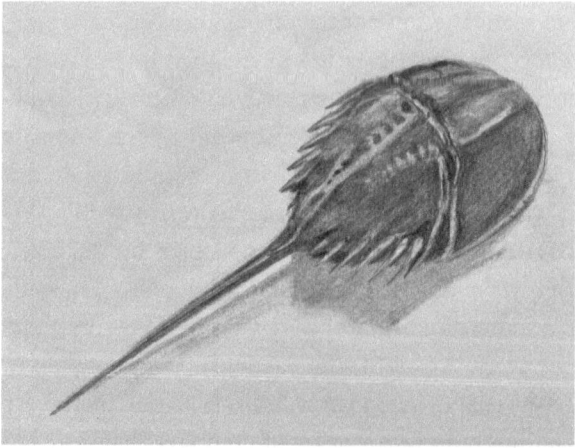

Horseshoe crabs have contributed immensely to research in medicine, pharmaceuticals, and physiology. Its blood contains primitive large circulating cells called amoebocytes from which a clotting agent is extracted called limulus amoebocyte lysate (LAL). Widely used by pharmaceutical companies to detect contamination by bacteria, LAL costs US$60,000 a gallon.

In Malaysia, I have encountered horseshoe crabs in the mangroves towards nightfall. They are seasonal visitors to the mangroves, invariable during high tide. At low tide, they may become stranded on the beach and can remain out of water for up to four days as long as they stay moist. To avoid drying out, horseshoe crabs often bury themselves in the sand or curl up into half its usual size to conserve water until the tide rises again.

Of all the marine species, horseshoe crabs have contributed most significantly to research in the fields of medicine, pharmaceuticals, and physiology. Its blood plays a vital role in human medicine as it contains primitive large circulating cells called amoebocytes from which a clotting agent is extracted called limulus amoebocyte lysate (LAL). This LAL has been widely used by pharmaceutical companies to detect contamination by bacteria. When the LAL comes in contact with bacterial contaminant called endotoxin, a clotting reaction occurs. This makes LAL most sought after by the medical and pharmaceutical industries for testing the sterility of vaccines, drugs, prosthetics, and other medical devices. Amoebocyte lysate has been employed to test drugs and vaccines since 1977. To date, there is no synthetic substitute for the LAL test. Today, horseshoe crabs are primarily farmed for their blood. Hundreds of thousands of them are milked of their copper-laden blood every year. The medical industry has come to rely heavily on this bluish-coloured blood. In recent times, the blood of horseshoe crabs is key to making Covid-19 vaccines safe, which is a crucial step in the manufacture of these vaccines. Tiny amounts of endotoxin contaminating the vaccines could spell disaster. Injectable drugs and prosthetics such as artificial limbs, hearts, knees, hips, and breast implants must also be completely sterile for use. Failure to ensure this can be deadly. Annually, pharmaceutical companies capture about half a million Atlantic horseshoe crabs, bleed them, and return them to the ocean. It is a time-consuming effort, and no wonder the resulting lysate can fetch a lucrative price as high as US$60,000 a gallon. I find it mind-boggling that we find ourselves existentially reliant on a primitive creature like the horseshoe crab for its unrefined and primordial body fluid.

5

Transit For World Travellers

In all the years of birdwatching, I always gave special attention to their colours and plumage patterns for correct identification to species level. If they were resting or perching on a branch, their voices and bodily antics would also help me to be more confident of what I was observing through my binoculars. Such conventional habits were perquisite to not only accurately identify birds but also learn about their relation to the habitats. Along the way, I found myself paying closer attention to the wholesomeness of nature as a side benefit to birdwatching. I soon realised that birds do not live in isolation despite their ability to fly and cover a wider range of habitations. Each species chooses to interact with numerous other animals and plants around them. In fact, they are also attracted to the non-living elements that surround them. The landscapes, geological formations, and local weather systems become just as important. I learned to understand the interconnections amongst the living and non-living things in nature by paying thoughtful attention to birds as an organism.

Indeed, birds are some of the best ambassadors from the natural world. There are approximately ten thousand species of birds spread across myriad habitats on all seven continents. Such innumerable species should not surprise anyone who is aware that birds are actually dinosaurs. In the view of palaeontologists today, birds are living dinosaurs. They are the only surviving lineage of dinosaurs that made it through the great extinction sixty-six million years ago. Tracing back evolutionary roots even further, researchers think all birds we see on planet Earth today evolved from a group of dinosaurs called theropods. They are all descendants of this one lineage of theropod dinosaurs. Other non-avian dinosaurs have long become extinct. A huge variety of bird species have evolved since then. What has contributed to this diversity of birds on planet Earth? Evolution and adaptation processes resulted in brightly coloured birds especially in the rainforests of the tropical regions. Having bright iridescent colours is generally crucial for the survival and continuance of bird species. Colourful feathers serve a number of different functions including soliciting mates through their plumage. They use colours and striking patterns to attract attention in courtship displays. They identify their own species through visible recognition of these colours and patterns. Colours also help birds avoid notice of their predators by having camouflaging plumage.

When I started birdwatching in a few coastal areas, it dawned on me that there was little thrill and excitement in spotting shorebirds including a number of species inhabiting the mangroves. It could not be further from the truth if I were to jump to a generalisation that birds found here are generally boring and uninteresting with respect to the colouration of their plumage. They are definitely less colourful. Their plumage is far from being equally striking and attractive as the rainforest birds. Monotonously dull, their plumage is restricted to variations of black, white, and grey. The most noticeably bright colours, if at all, appear on their bills and legs.

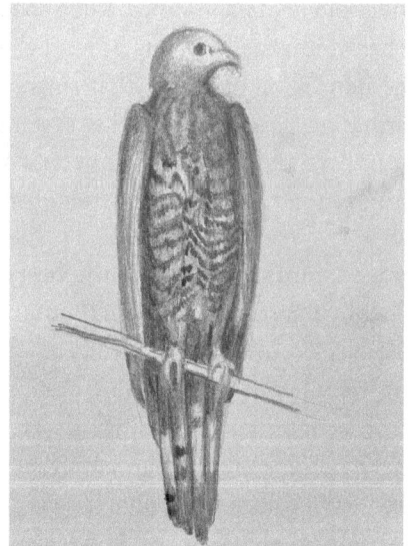

Birds of the tropical rainforest put to shame those species in temperate regions, which are often monotonously brown and grey in appearance. The plumage of tropical birds tends to be more colourful and used to attract mates or warn away competitors. Some tropical species may have dull plumage, which can be more important in providing camouflage from predators.

Birdwatchers and nature enthusiasts are forever on the lookout for awesome places to pursue and delight their passion about wildlife and nature. In Peninsular Malaysia, there is a gazetted forest reserve known as Tanjung Tuan, located at the border of Negeri Sembilan and Melaka, which has over the years emerged as a go-to natural site to watch migratory birds. Every March, thousands of huge birds called raptors leave their wintering sites in Indonesia and Australia to head north to their breeding sites in China, Japan, Korea, Mongolia, and Siberia. Raptor is a generic term for all birds of prey with strong bills and large talons. Without exception, all raptors are seriously carnivorous. Their flight path to the final breeding destinations can be up to ten thousand kilometres, and migrations usually take up to seventy days. This prodigious highway in the sky has been dubbed the East Asian-Australasian Flyway. During this northbound migration, thousands of raptors can be seen crossing the Straits of Malacca from Sumatra to Peninsular Malaysia. Estimates from Tanjung Tuan birdwatchers are one hundred thousand migratory birds of at least twenty bird species use this route annually. They are primarily raptors from eleven species, which in total might account to as many as forty thousand individuals per season. The bulk of the migratory birds using this Asian-Australian Flyway are from ten other species involving more than seventy thousand individuals. Oriental honey buzzard, *Pernis ptilorhyncus*, makes up more than 80 per cent of the flight, followed by black baza, *Aviceda leuphotes*, at around 9 per cent, and 3 per cent Chinese goshawk, *Accipiter soloensis*. Other world travellers often sighted at Tanjung Tuan include black-shouldered kites, *Elanus caeruleus*; Japanese sparrowhawk, *Accipiter gularis*; greater spotted eagle, *Clanga clanga*; and grey-faced buzzard, *Butastur indicus*. That is quite a crowd.

Raptors are birds of prey that live on carrions or seized prey. Their diet,

consisting entirely of meat, sees this predator evolving and adapting to become one of the most remarkable hunters from the air.

Raptors are set apart from other birds in their possession of powerful feet armed with sharp talons used to seize prey, making their hunting success highly rated in the animal world.

Raptors are predators that occupy the top of the food chain. Interestingly, if their population is reduced or threatened in the wild, we can conclude that other animals in that ecosystem are also at risk.

Most migration appeared to take place during days when wind direction was blowing from southwest and west. Tanjung Tuan Forest Reserve in Melaka happens to be ideally situated in the middle of this route. Tanjung Tuan is a hilly headland that juts into the sea, making it the narrowest part of the Straits of Melaka between Sumatra and Peninsular Malaysia. Such physical features and relatively dry landscape are much preferred by migratory raptors, which do not like crossing large bodies of water. Generally, large heavy birds prefer land routes, which have thermal or hot air currents they can take advantage of to lift them up and remain aloft for hundreds of kilometres. This way, little effort and energy is required. As they cross from Sumatra, the birds will have slowly lost altitude with the thermal air cooling. Then all they need to do is to flap their wings to stay airborne. After crossing the Straits of Melaka, the birds have expended most of their energy and start to land at Tanjung Tuan. They will be ready to rise again when the thermal air is regenerated as the air current warms up by mid-morning the next day. This yearly event called Raptor Watch has become the most awaited spectacle of nature by birdwatchers throughout the region. It has become an international event limited not only to the bird migrators but also to the avid birdwatchers.

Other migratory birds, however, prefer wetlands and mangroves as their crucial stopover sites. Unlike raptors, seabirds including gannets, boobies, gulls, terns, and petrels prefer taking a much-needed rest after their long flight in places with vast water bodies. This preference enables them to migrate very long distances because they can rest on water as well as eat. Also commonly found on beaches or inland mudflats of mangroves are shorebirds including wading birds such as plovers, sandpipers, and snipes. They, too, need to migrate during the breeding season. But long, non-stop flying is no feat for some of these shorebird species because they cannot remain afloat on the sea surface like seabirds. Shorebirds are not sufficiently skilled at fishing in the open sea like seabirds. And so the migratory shorebirds choose to break their journey into shorter sections of a few hundred kilometres each. Several mangrove forests in the tropics have become their saviour in overcoming the otherwise long arduous journeys.

In 2010, I retired to immigrate to my wife's hometown of Invercargill in New Zealand. It was a hard decision to make after more than four decades building my career as a teacher and researcher in many parts of Malaysia. What would I do with the limitless free time at hand? It turned out to be a good decision after all as I began to get used to killing time through writing and drawing. I was totally enjoying my retirement until I received a startling call from the Sabah Foundation, a Malaysian state organisation established in 1966 with the primary aim of promoting educational and economic opportunities for the people of Sabah. I was asked if I would be interested to return to establish a new university. I consulted my wife to seek her views first because it would mean leaving our newly renovated house empty for at least three years. Knowing how much I loved Sabah and have always been looking for another opportunity to return and contribute to the education of Sabah youth, she left the decision to me. That was a no-brainer. I answered in the positive, which resulted in me becoming the founding vice chancellor of University College Sabah Foundation (UCSF).

In Kota Kinabalu, the state capital of Sabah, I was living alone in a beautiful apartment facing the sea. A mere thirty-minute walk from where I was staying was Kota Kinabalu Wetland. In the 1980s, these twenty-four hectares of mangrove forest was considered a useless mangrove swamp area only fit for dumping municipal waste. At the time, I thought the entire area would end up as yet another landfill site like many mangrove areas in the region. But today, the mangrove reserve is to be left untouched in perpetuity. In 2016, Kota Kinabalu Wetland, sited in the heart of Kota Kinabalu City, was gazetted as Malaysia's seventh Wetland of International Importance under the Ramsar Convention on Wetlands, an intergovernmental treaty for the conservation and wise use of wetlands. I am extremely pleased to learn that, today, Kota Kinabalu Wetland serves as a foremost sanctuary and feeding ground for many species of resident and migratory birds. Numerous species of migratory birds from Northern Asia are known to use the site as a transit in their flight across the globe from September to April. Some in fact stay to breed, not continuing to complete their long migratory journey with the rest of the flock. The more common winter visitors to Borneo include little egrets, striated herons, white-winged terns, and black-crowned night herons. Sometimes, the common moorhens, cinnamon bitterns, and common redshanks are also spotted there albeit rarely. Common residents like the white-breasted waterhens and marsh sandpipers are seen all year round.

Tens of millions of wondrous migrating birds use the tropical wetlands as a transit on their flight heading for warmer weather and breeding grounds. Some travel relatively short distances, while others cover thousands of kilometres. These wetlands including mangroves, mudflats, lakes, lagoons, and rivers are under increasing threat of drying out and pollution, leaving migratory birds at growing risk of dying.

Not all birds need to perform the arduous long flight in search of breeding grounds. A number of shorebird species are migratory too, but they don't have to make the flight across the globe. They much prefer to stay around to breed as they are not equipped to take on the long migratory journey.

On one of my many trips there, I had the wonderful experience of seeing literally thousands of cattle egrets in transit. I have always wondered how this particular bird species has conquered the world. Cattle egret, *Bubulcus ibis*, has become so widespread in its distribution throughout the world that scientists are still grappling to explain how and when they did so. As a child growing up in a village located close to rice fields, I have always associated this heron-related bird with the water buffaloes. The images of the snowy white plumage of cattle egrets against the background of black-bodied water buffaloes are impossible to erase from my childhood memories. Now I no longer marvel at such a familiar sight. Instead, the question of how they have spread across the world so quickly has intrigued me more. In the past 150 years, they have been breeding in nearly every continent on earth. Just how and why remain a nagging mystery to bird enthusiasts.

Cattle egret, *Bubulcus ibis*, is widespread across all continents of the world. One of the mysteries in ornithology is how this particular bird species has conquered the world. Scientists are still grappling to explain how and when they did so.

True to its name, the cattle egret is invariably seen feeding and trailing after herds of cattle or other large mammals like water buffaloes. They are insectivores, feeding on both land and aquatic insects found inhabiting the wetlands. This is quite a departure from the feeding habits of other heron and egret species, which usually feed on fish and crustaceans. Only occasionally, cattle egrets supplement their diet with more typical heron cuisines like frogs, tadpoles, crayfish, lizards, molluscs, and even small birds or mammals. That still does not explain why cattle egrets prefer to be around cattle like water buffaloes in tropical areas. Birders in Africa have also associated cattle egrets with other herding mammals like wildebeest and elephants. As recently as 150 years ago, one had to travel to tropical Africa or the southern Iberian Peninsula to see a cattle egret. Then in the mid-1900s, seasonal small populations of cattle egret began to establish northwards throughout Europe. They also spread southwards into South Africa, and soon they began to cross the Atlantic. They were first documented in late 1880s on the border between Guyana and Suriname, suggesting that was the time cattle egrets arrived on their first flight across the Atlantic. But how they did it remains a mystery.

The cattle egrets and grazing cattle find themselves in a relationship called commensalism. They are always seen feeding close to where the cattle are grazing. As the cattle moves, insects are stirred up and flushed out from the grassy vegetation, and egrets are there ready for a feast without much effort.

Some modelling studies can provide a clearer picture how these first cattle egrets were able to colonise the New World from the African coast. One suggestion was their long flight across the Atlantic in less than a week could have been aided by the annual trade winds between March and April. By the late 1940s, large populations of cattle egret were firmly established on the northern end of the continent. As soon as the 1970s, they were found breeding as far south as Chile. In the United States, the species was first recorded in Florida as recently as in 1941. Their breeding population was documented in 1953 when they began to spread along the coastlines, pushing gradually inland, and by 1962, they were breeding in Canada. Around the Atlantic, the species continued to broaden its range into India and Southeast Asia. The first cattle egrets arrived in Australia in the 1940s, presumably from Asia by way of Indonesia and Melanesia. By the early 1960s, they had hopped across the sea to New Zealand. Such is our current guesstimate about the routes and migration patterns of cattle egrets to self-populate and achieve global domination—peacefully and quickly too. But to me, the bigger mystery is how they manage to keep this long tradition of flying long distances to and from their breeding grounds each year. Their ability to fly long distances is mind-boggling. That to me is an amazing story how a winged creature could have swiftly come, seen, and conquered the world!

Many migratory bird species fly mainly during the night; some during the daytime, and others both night and day. Nocturnal migrations are said to be more advantageous since the birds can maximise foraging during the day and optimise energy expended on long flights under cooler night temperatures. After all, a flight of thousands of kilometres is a high-octane endeavour.

Bird migration is a natural process. Bird species that live in regions with extreme temperature changes from summer to winter survive by moving with the seasons to more favourable conditions elsewhere. As a rule, they migrate to move from areas of low or decreasing resources to areas of high or increasing resources. The two primary resources being sought are food and nesting locations. As winter approaches and the availability of insects and other food drops, they have to set off on mighty migrations to find food and better places to give birth and breed. Some of these journeys are long and difficult. Many make record-breaking journeys across continents and oceans. Each winter, birds of northern and southern polar waters would embark on a massive migration to breed in warm locations near the equator. They brave the long and tiring journey across oceans in massive numbers annually. Tracking studies revealed migrating birds are able to make annual journeys of about seventy-one thousand kilometres. Depending on the species and the prevailing winds, this means they are travelling at a speed within the range of twenty-five to ninety kilometres per hour. At these rates, migratory birds typically fly from twenty-five to nine hundred kilometres or more each day. On record, the Arctic tern is the world's champion long-distance migrant. Their migration over thousands of miles provides proof of their great stamina and skill.

There are still mysteries about bird migration and their many different migration patterns. Our knowledge about the destinations and routes of bird migrations has accumulated from using tracking methods of modern technologies. But how these migratory birds find their destinations halfway across the world is still an open question. We still marvel at the means such small creatures can travel so far year after year as a matter of routine. There are still more questions than answers. How do they accomplish it? During this long arduous flight, migratory birds cannot afford to lose their way. They constantly need to know which direction to fly. Indeed, they all seem to know to navigate correctly, and rarely do any go astray. Did the young ones learn this from their elders, or were they born with this incredible navigational skill?

At all times, their navigational skill must be impeccable so as not to lose their way. An error leading to missing their breeding destination would spell disaster for the species.

Evidence exists to suggest birds can understand the magnetic fields that surround the earth. These magnetic fields run from the North to South Poles. This, however, doesn't explain their navigational wonder. There are two basic processes in true navigation: piloting and orientation. For piloting, the birds can use the ordinary sight, smell, or sound like we humans do. Piloting, therefore, can rely on specific landmarks found in the vast oceans or terrestrial landscapes. A red buoy in the vast blue oceans may be used as a landmark to indicate a good fishing ground. The sight of a huge shimmering wetland in the hinterland may also be used to pilot migrating birds. Interestingly, landmarks need not be restricted to things they can see. Odour and sounds can be landmarks too. The smell of rotting carcasses can also guide them in piloting to specific destinations. The sound of frogs croaking in water bodies is another. In theory, it is also possible for infra-sounds to be generated by winds blowing over mountainous regions or huge waves pounding on shores. These sounds can be heard for thousands of kilometres. They, too, may become landmarks for migratory birds to pilot themselves towards their destinations.

Besides piloting, they also have to orientate themselves during flight. Orientation is the ability to position themselves. This is akin to our setting a fixed course on our compass before setting off on or during a journey. For instance, we necessarily orientate ourselves before we set off on a sea voyage or going on a jungle walk because our destination is not in plain sight. We conventionally rely on the north-south-east-west compass to orientate. For animals, the positions of the stars and the moon may serve as their compass on long-distance travel where specific destination are yet to be visible. Flying during the day, the sun can serve as a compass. The migratory birds need to know the accurate time of the day when migrating. For instance, in the northern hemisphere, whatever the season of the year, the sun is always moving south at noon. A migrating bird, therefore, needs to maintain a course by continually changing the angle between its flight path and the sun. Failing to do this, the bird will find itself flying in a circle. The sun cannot be the only compass to maintain their orientation because migrating birds also keep a good course on cloudy days when the sun is hidden. At night, migratory birds are also able to orientate. How does this happen in darkness? Most migratory species have been shown to be aware of the rotation of the stars around them. They are able to identify the one stationary star around which the other stars move. This is called the Pole Star. As soon as they identify this Pole Star, they orientate their flight away from it at all times. This would eventually take them south.

An increasingly acceptable theory is they use a pervasive pair-based compass to 'see' Earth's magnetic field. This allows them to embark on long migrations in great numbers without getting lost. Experiments have shown birds can orientate when moving through a magnetic field. They fail to fly in the correct direction when magnetic fields are experimentally changed or altered by an electromagnet. They are able to detect these changes by means of a highly magnetic mineral called magnetite in their brains. Experts believe this mineral may help birds detect Earth's magnetic field and use it to guide them south. In fact, similar crystals of magnetite have been shown to exist in human beings as well as in bacteria. How these tiny magnetic crystals work is still a big mystery.

Perhaps studying how migratory birds navigate by sun, stars, and magnetic fields could potentially reveal many more elusive mysteries in animal behaviours, including us humans. One that immediately comes to mind is how we often blame it on the full moon whenever someone has gone totally berserk. Behavioural changes in humans have long been said to be influenced by the full moon above. Human madness and the moon have had a long causal-effect association. The term *lunatic* is derived from *lunar*, which is Latin for 'moon'. Perhaps this is why I often become rather irrational during the full moon. *How beseechingly interesting*, I thought.

All is not well in bird migration. There are disadvantages in this annual phenomenon too. It could spell doom for some predators. In some ecosystems where birds are the primary food source, predators could die of starvation because of the shortage of food when birds in their habitats migrate. For the migrating birds, they, too, face several risks during their long journey. Enormous numbers of birds can die on flight. This could be out of fatigue or killed by diseases they might be carrying. To avoid predators, flying in the dark of the night could escape notice by land predators. Bird species of smaller sizes seem to migrate during the night. Many bird predators are more active during the day, and so migrating at night makes small birds less vulnerable to predation. The migratory flight is often accomplished in one stretch without stopping to rest their flight muscles. They don't even rest to feed and replenish their energy. This indicates they are able to maximise the availability of energy throughout the long-haul journey. Flying by night offers some clear advantages. The skies are often less turbulent during the night, allowing for easier flight and ability to remain on course. Air temperatures are typically cooler at night. They can save their metabolic fuel better during migration, which is a high-energy activity. Indeed, bird migration is a prodigious feat of endurance often over hostile environments. Even if birds have evolved and adapted to obtaining the necessary energy to do so, is it worth all the hazards and efforts?

I decided to explore further towards the less vegetated stretch of the wetland along the water's edge. I felt like a quiet evening stroll along the beach. The orange sinking sun set the entire landscape aglow. I sank my feet in the soft muddy flat cautiously not to completely drench the heels of my shoes. Against the fiery panorama of the sunset, I caught sight of a solitary striated heron, *Butorides striata*, also known as little heron or green-backed heron. Peering through my binoculars, I confirmed that it was a small heron not more than forty-five centimetres in height with a blue-grey back and wings. Its whitish underparts caught some of the glow from the setting sun. Around its black head, a dark line extended from the bill to just underneath the eyes. As typically the case with shorebirds, the only bright colouration is its pair of short yellow legs and yellow bill. It was foraging along the water's edge, prodding along in almost Zen-like calmness. I stopped to venerate its incredibly calm behaviour. The bird would suddenly ambush its prey with its sharp pointed bill. It stopped to stand still at the water's edge for some time in wait to ambush its prey, which included small fish, frogs, and aquatic insects. It also perched and waited on a branch or stilt root over the water, tucking in its neck, crouching in a low forward position. It flicked its crest up and down as it waited patiently. Little heron is noted for its amazing fishing ingenuity. Cleverly, little heron sometimes uses bait to lure victims closer to the water's edge. When the fish approaches to end up swimming around its partially submerged legs, it would give the final blow on the victim in a most timely fashion. In the natural world, the heron has been observed to drop a feather or leaf carefully and deliberately on the water surface. As the fish comes to investigate, the bird would swiftly pick it out of the water. I have seen a video recording of such fishing wizardry in little heron found living near water bodies in urban areas. Under such circumstances, the heron would improvise its bait using food scraps in public parks. Bread left on benches or discarded in rubbish bins, for instance, is used as bait, not necessarily something from nature.

Striated herons seem to have acquired the three requirements in fishing just like humans. Any expert angler would tell you that to catch a fish, you need the right bait, the perfect spot by the water's edge, and patience.

Striated herons have mastered these fishing techniques in their natural environment as well as in urban settings close to human habitation.

The vast mudflat zone tends to expose vulnerable creatures like the fiddler crabs and hermit crabs to a whole range of predators. Here, the 'Eat or be eaten' axiom becomes more devious and prevalent. I looked over my shoulder to assess the extent of this predator-prey relationship that might be operating in this part of the wetland. I spotted a working party of common sandpipers, *Actitis hypoleucos*, not far from where I was standing. They could be easily identified by their habit of 'teetering', constantly bobbing their head and tail. They seemed to be always busy feeding on the mudflat, never in a hurry. Indeed, the success of this bird appeared to be related to their feeding habit. A generalist in their feeding behaviour and having no special requirement in their diet, they eat a wide range and variety of food, from minute invertebrates to crabs, bivalves, worms, insects, spiders, and centipedes. Therefore, sandpipers are less sensitive to changes in their habitat, particularly seasonal changes that could affect their food availability.

Sandpipers display a remarkable sense of movement despite its constant teetering and bobbing of the tail up and down. Far from being anxious or uneasy, they are often seen solitarily wading in the shallow pools of the mangroves dining on crabs, insects, and other sea creatures.

After an engrossing few hours in the hot and humid air of the magical mangroves, I was exhausted and ready to head back to the air-conditioned condominium. But as I approached the entrance of the reserve, I caught sight of a silhouette of a bird I have seen on a number of occasions along rivers in the interiors of Sabah. But this time, it was close to human habitation only a few minutes' drive from the city. *Strange*, I thought. Spying through my binoculars, I immediately recognised an Oriental darter, *Anhinga melanogaster*, commonly known as snakebird. It is a waterbird commonly found in wetlands and along the rivers of the tropical rainforests. Taxonomically, Oriental darter is a cormorant-like species most recognisable by its white lateral stripe running along its very long neck. It is a fish eater and often swims with only the neck above water. This gives a snake-like appearance when they swim with their bodies submerged, hence the common name snakebird. Kota Kinabalu Wetland must have attracted this usually inland species because of the abundance of fish found in the lagoon at that time of the year.

The Oriental darter (*Anhinga melanogaster*) is equipped with a long, slender neck and straight, pointed bill for spearing fish while submerged in water. It is built to absorb the force as its face hits the water beak first by tucking the wings to the sides, turning its entire body into a feather-covered aquatic missile. The momentum carries it deep into the water where it uses its feet to swim.

The skeletal and muscular structures of darter birds have evolved and adapted to make this species an excellent fisher. Its fifth to seventh vertebrae are attached to muscles, enabling it to project the bill forward like a spear. Its sharply pointed bill is skilfully employed to seize prey when they dive for fish in the water. This why they obtained the name darter. They have been seen to bring their catch to the surface, toss the stabbed prey into the air, catch it with the bill just to eventually swallow it whole—a juggling act one could never forget once you've seen it. Their diet comes from a variety of amphibians such as frogs, newts, and reptiles such as snakes and turtles. Sometimes, invertebrates such as insects, shrimp, and molluscs can feature on the menu as well.

A perching Oriental darter bird after its hunting trip is also a sight that if you see once remains in your memory bank forever. After diving for fish in the river or mangrove, its plumage needs to be adequately dried so as to fly smoothly again. This is done by finding a dead branch clearly exposed to the sun, perch on it, and spread its wings wide. A long sunbathing time is required fort their wet feathers to dry and become silky again. It the wild, it definitely looks unusual and of no purpose for a bird to have its wings spread out almost fully for a long time without taking off on flight.

Outside, the majestic peak of Mount Kinabalu is visible from the sixteenth floor of my condominium. The familiar outline of this iconic mountain began to gradually fade in the dark stillness of the moonlit evening. Tired, I placed my head gently on the pillow. The many chanced encounters of the day consumed my thoughts. An awesome experience and what a feeling! Granted I wasn't anywhere near *Star Trek*'s mission of 'To boldly go where no man has gone before', I felt quite contented and accomplished. There is no simple description to the aura of satisfaction and the unending excitement about what I had seen and experienced on my many trips to the Kota Kinabalu Wetland throughout my close to two years' living alone in Sabah. It definitely served to calm my delirious brain, thinking of ways to start a new university on a firm footing. Finally, even the constant sounds of traffic zipping along the Teluk Likas Highway below couldn't keep me from having a good night's rest to start another day at the office the following day.

6

Getting About

I was on air being interviewed in a radio programme about career choices for graduates. The host asked what I considered most fulfilling after being an academician and research scientist for more than four decades. Without much thought, I unhesitatingly replied, 'Travel!' Although answered hurriedly, that was not subconscious. All the travel I have done and learned throughout my career have almost made me as complete as I would like to be—definitely beyond my dreams and expectations. I have seen and experienced living in many places through travelling since a young man pursuing my education and career. I've learned and unlearned many things. For a good part of my life, I found myself moving to places outside my country of birth as a student in New Zealand, Thailand, and the United States. Upon graduation with a doctoral degree, I spent time building my career in all the three regions of my own country: Peninsular Malaysia, Sabah, and Sarawak. For my post-doctorate and sabbatical leave, I spent short stints doing research in Indianapolis, Ann Arbor, Michigan, USA, and Northeast Wales UK. I have also attended scores of scientific meetings and conferences in countries across almost every continent. Throughout my career, I have also seized the many opportunities to travel and explore a multitude of natural ecosystems in many parts of the world. They were all trips to satisfy my unending curiosity aimed at learning more about nature and culture. But the underlying benefits to all these fun trips were also to improve my well-being as a family provider. It results in specifics like having a strong and satisfying career, having a nice house to live in, being able to put food on the table, adequately providing and raising a family, and ensuring a good future for my children. In retrospect, travel somehow comes in as an inevitable overarching requirement to all of that.

I was in Johor Bharu to present a paper at the International Youth Congress on Biodiversity Conservation in 2006. An old acquaintance from the University Technology Malaysia (UTM) suggested we should visit Tanjong

Piai, Johor National Park, to which I unhesitatingly agreed. The national park covers 926 hectares of mangroves and mudflats. I have come across a promotional claim in an ecotourism brochure that Tanjung Piai, Johor National Park, is located at the southernmost tip of Peninsular Malaysia, which also makes it as far south as you can go on the continent of Asia or Eurasia. That sounds so cool!

A network of wooden boardwalks was constructed to facilitate access to different areas of the reserve without having to negotiate the way through the inhospitable prop roots and muddy floors of the mangroves. From the platform, I was busy clicking away getting awesome snapshots of mudskippers, crabs, and cockles splish-splashing about on the grey mud below. There are observation towers that have been specially constructed for birdwatchers and nature lovers to obtain a better view of the ecosystem and its living constituents. I cautiously climbed up to the observation platform to catch a view out to sea where ocean liners, petroleum tankers, and other vessels were queuing up to enter Singapore via the Johor Strait. I was thrilled to catch a glimpse of one of the world's busiest shipping lanes from up there.

Training my eyes of the vastness of the reserve, I noticed a few species of birds of prey circling the skyway. There is nothing more awe-inspiring than to watch a bird of prey hovering and circling across a clear blue sky and suddenly swooping in for a kill on its prey. The lesser fish eagle, *Haliaeetus humilis*, just gripped and pulled up a slippery lunch from the sea using its curved talons. Birds of prey are truly doyens of all birds in the sky. They are capable of swooping, soaring, and hovering in incredible displays of flying skill. I have seen a pair of sea eagles fighting in mid-air to snatch a sea snake, *Laticauda colubrine*, from each other's grips. They were doing all sorts of acrobatic flights twisting and tumbling in the air with their claws grabbing at each end of the prey. When one lost the grip, it would glide gracefully and circle around before making another swoop in an attempt to grab the snake again. A mid-air battle between a pair of sea eagles is one of the most dramatic scenes of nature in action that remains etched in my memory.

White-bellied sea eagles are masters of the sea and sky. They come equipped with a slew of adaptations that make them maestro of high-speed skydiving. Spotting a fish in the water from a height of up to thirty metres, they can swoop down to pick up the target with a dive exceeding eighty kilometres per hour.

Unequivocally a master in flight, the prowess and manoeuvrability of birds of prey are unrivalled. They owe this to their strong lightweight bones, powerful breast muscles, and efficient circulatory system. Most impressive is their ability to perform fast-moving acrobatics while in flight. Evolution has finely engineered all birds of prey to suit their lifestyle. The degree of lift, thrust, and movement all are primarily determined by the shape and size of their wings and tail. Short wings enable fast acceleration, while long, thin wings enable graceful gliding and remaining airborne longer without tiring. In combination, they give this epic predator the speed required for pursuit of their prey. In mid-air, birds of prey have been known to clock a speed of about three hundred kilometres per hour. They would, however, be almost handicapped and at a loss without their claws and talons. These are vital weapons used to seize their prey. They give the bird a powerful grasp as it swoops down with its toes and sharp talons ready to close around the victim. Its feet can quickly lock in and continuously hold the prey for hours without tiring.

There is more to this. This masterful skill in pouncing on its prey can further be attributed to its pair of eyes positioned towards the front of the head, not at the sides as with other birds. As such, it gives them binocular vision with intersecting fields of view, enabling good perception of depth. Not having eyes set wide apart on the sides of their head like other birds might give them the disadvantage of not being able to locate prey or see danger from all around. But that becomes less important in birds of prey. In flight, these birds do not have to keep a look out for danger. Instead, they need to improve focus and precision on locating prey in front. Their eyesight is determined to be four times sharper than that of humans. Their retina has thousands of photoreceptors all connected to send information to the brain. The information gathered by both eyes is then combined and processed to achieve a perfect focus and to judge distances in a fraction of a second.

Binocular vision enables birds of prey to maintain visual focus on an object with both eyes, creating a single visual image. This is absolutely crucial in eliminating distortions in depth perception and in visual

measurement of distance. Eagles are able to rotate their head, extending their visual field even more than humans.

The agility of movement by the birds of prey totally impressed me. I was moved to reflect on the value of locomotion in the animal kingdom. Travel and getting about for animals is purposive and a matter of survival— definitely not a spur-of-the-moment activity like I used to travel and see places at the slightest of excuses. Getting about is not a unique prerequisite for humans but imperative in the animal kingdom. Highly successful species are those able to move around. Over millions of years, animal species have evolved from being sedentary to mobile. Sedentary species inhabit the same locality throughout life, whereas mobile species are capable of moving themselves in a variety of ways for different purposes to survive. In the coral reefs, sedentary organisms include sponges, hydroids, and coral polyps. In the mangroves within the intertidal zones, sedentary animals are represented by tube worms, barnacles, and some bivalves. These are considered primitive life forms that have not evolved over billions of years to afford them the survival advantage.

The mangrove forests appear to be the zone where the evolutionary transition between sedentary and mobile species could have taken place. Locomotion, the ability to move from place to place, appears more varied and sophisticated in the mangroves compared to the adjacent coral reefs out at sea. All the three types of locomotion—flying, swimming, and terrestrial—are prevalent amongst animals inhabiting the mangroves. Undoubtedly, the ability to move from place to place is one of the earlier products of evolution adapted by mangrove species for survival on land. Taking advantage of the mobility accorded by locomotion, the animals here are able to perform everything from foraging for food to finding mating partners and escaping from predators.

Flying seems the ultimate mode of getting about quickly and safely. Active flight through the air is only enjoyed by select groups of animals including insects, birds, and one mammal species, the bats. Active flight should not be confused with passive flight. The ability of some animals to traverse through air need not always be active flight. Some merely glide downwards from taller to lower heights. These animals include flying squirrels, flying frogs, flying snakes, and flying lizards of the tropical rainforests. Flying is a misnomer here. They should be referred to as gliding animals.

Animal species that have evolved to truly fly have special adaptations enabling them to not only defy and overcome the force of gravity but also consume as little energy as possible staying aloft. To accomplish the latter criteria, they possess minimal body mass. A case in point is the birds' bone structure. Unlike the tough and heavy bones of humans, birds have almost hollow bones, mostly air-filled regions. Such anatomy gives the birds overall mass unbelievably low—light enough to float in air with little uplift pressure. Evolution also has not given birds any teeth, which helps to further lessen its overall body weight. Interestingly also, birds do not have a urinary bladder included as part of their secretory system. There is a plausible reason for this. They do not have to fly around carrying a bladder filled with urine, avoiding the burden of hauling along a heavy 'water balloon' when in flight.

Most importantly is the ability of birds to overcome gravity. For this, birds have evolved wing shapes that help them to defy gravity with as little energy as possible, an evolutionary ingenuity. The unique shape of their wings allows them to alter the air currents around them as they fly through the air. It is all about reducing the drag force as they fly. The soft light feathers huddle together uniformly to form a torpedo-shaped body as they cut through the air seamlessly with as little energy as possible. Staying afloat and travelling to places seem effortless—a piece of cake!

Designed to fly, the vastness and openness of mangrove forests become the favourite hunting grounds for birds. All around in this unique mangrove ecosystem, one can witness how animal species gifted with the remarkable feat of flying can become supremely successful. I spotted another bird of prey, the majestic brahminy kite, *Haliastur indus*, gracefully encircling above against the cloudless blue sky.

The design of bird wings helps birds of prey to defy gravity with as little energy as possible. The unique shapes and anatomy of their wings alter the air currents around the body, causing a reduction of the drag force as they fly through the air.

Brahminy kite, *Haliastur indus*, a handsome white-headed rusty-red raptor, has a strong spiritual association with Hinduism and Indian mythology. Referred to as Garuda, this most venerated bird is the

sacred bird of Vishnu in Hindu culture. It derives its name from the term *Brahmin*, a varna or special class in Hinduism to comprise only intellectuals, priests, teachers, ayurvedic physicians, and protectors of sacred learning across generations.

Animals that do not fly have evolved means to move on land in equally beguiling ways. Anyone who has had the opportunity to handle a snake with bare hands would remember not just being unsure and nervous but also how squirmy it feels. Its skin along the entire body is slippery. One gets the impression that a snake appears to advance forward smoothly because their skin is overall slippery—that it swimmingly moves free of bumps and friction against the surface it was slithering on. Such assumption is furthest from the truth. Snakes in fact depend on friction to move. They crawl by contracting the muscles that run along their body with the aim of creating friction between their underside against the ground or rough bark of trees. To achieve this, they possess scales, which can create maximal friction by rubbing and catching on the minute bumps present on rough surfaces. That is why we feel the length of their body smooth only when we stroke the animal towards its tail. But stroking a snake along the body towards its head will give you the rough sensation like rubbing your palm on sandpaper.

When we are asked to draw a snake, we invariably draw its stereotypic S-shaped body. Rarely do we picture them crawling effortlessly in a straight line. But to move forward calmly in a straight line is in fact more efficient than to wiggle waves along its slender body as they appear to be scuttling in a hurry. Snakes have been shown to create twice as much friction when moving forward straight than wiggling sideways. This is why a number of heavily built constrictor species like the pythons, anacondas, and boas move forward on the forest floor in a straight line. They do not have to wiggle to move forward but propel themselves forward like a train through a tunnel.

Snakes don't have to wiggle sideways to propel themselves forward. In fact, snakes can move forward in a straight line twice as fast because they rely on the friction between their underbelly and the surface they are travelling caused by muscle activity and loose belly skin, which has scales facing backwards towards the tail. Heavily built constrictor species like the pythons, anacondas, and boas do not wiggle but move forward on the forest floor in a straight line.

This ability to travel in a straight line is achieved by the coordination of muscle activity and their skin movement. When the snake moves forward, the loose skin on its belly flexes substantially more than the skin over its ribcage and back. In a similar fashion to the treads on vehicle tyres, the belly scales function in providing traction with the ground. The muscles pull the snake's internal skeletal structure forward in an undulating pattern. Because the muscles below are sequentially activated from the head towards the tail, the forward movement of the whole snake appears fluid and seamless to our eyes even when it is moving quickly. Such synchronised contraction of underside muscles along the loose and more flexible belly skin is able to carry itself forward in a straight line without having to wiggle and bend its spine. This ability is thought to have an added advantage for bulky species of snakes. It makes more sense because, evolutionarily speaking, snakes evolved from burrowing ancestors. The ability to straighten its long body would make fitting through narrow holes or channels easier than having to bend its body and push against a solid surface.

One has not seen the magic of mangroves without an amazing encounter with a fish that can walk on land—another adaptation to survive in the mangrove environment that is simply bizarre. A type of amphibious fish, the mudskippers are able to spend time out of water as normal parts of their life. As the name implies, mudskippers move by skipping or hopping across the mud surface. There are thirty-two species of mudskippers. All are known for their unusual appearance and ability to survive both in and out of water. They can grow up to thirty centimetres in length, and most are brownish green in colour, ranging anywhere from dark to light. During the mating season, the males are observed to develop brightly coloured spots to attract females. The spots can be bright iridescent red, green, and even blue.

Amphibious fish including the mudskippers have attracted much attention from scientists undertaking theoretical research in evolution. The adaptive evolution of animals from aquatic to land environment has eluded scientists for a long time. Amphibious fish like the mudskipper are thought to serve as models in understanding the critical changes that could have been manifested during the transition of living animals from water to a terrestrial lifestyle. Could mudskippers provide some clues to that long-sought missing link in the evolution of our fish ancestors from water to a life capable of walking on four limbs on land?

In the mangroves, mudskippers spend part of their lives out of water to feed, mate, and avoid capture by terrestrial predators. To have successfully lived in the uniquely harsh environment of the mangroves, mudskippers owe that to some fascinating modifications in their morphology and physiology. They are usually found around the root systems of the mangroves or on the open mudflats. Out of the water, they are well adapted to remain on land after the tide has gone out. Although they have no special organs for breathing air, they can absorb oxygen through their skin and the lining of their mouth as long as they stay moist. In addition, they can retain bubbles of water inside their gill chambers, allowing them to carry on breathing through their gills while on land. This enables them to remain on land for up to two days at a time.

Mudskippers are found outside their burrows around the mangrove roots or in the open mudflats. They are well adapted to be out of the water after the tide has long ebbed and can continue foraging or seeking mates. This is made possible through their side pectoral fins located at the frontal region, which can function similarly to the crutches humans use to get around when their legs are incapacitated.

A mudskipper's eyes protrude from the top of its flat head. This pair of goggle eyes strangely situated on the head is a clever adaptation to their amphibious way of life. Strictly aqueous fish are likely to have myopic vision in air when out of water. However, mudskippers seem to possess good vision on land. Being endowed with a pair of close-set and moveable bulging eyes gives the mudskipper a more accurate and panoramic vision when out of the water. This could be most important for escape from predators such as shorebirds wading on the mudflats and birds of prey swooping from the sky.

The most noticeable feature on mudskippers is their side pectoral fins located in the frontal region under their elongated body. These fins function similarly to legs in that they allow the mudskipper to move on land from place to place. The only other fish I am aware of that can survive out of water for short periods is the African lungfish. Like the mudskipper, the lungfish uses its slender fins to 'walk' on land. The lungfish is also thought to be a vital link between aquatic and land-dwelling animals. The mobility of mudskippers on land is fascinating to watch as it uses the stubby pectoral fins like humans use crutches. This robust pair of fins enable the fish to push themselves off the ground and 'walk' forward. They allow the mudskippers to roam across muddy surfaces looking for food. This novel locomotion even gives them the ability to climb trees and low branches. In water, mudskippers use their pectoral fins, caudal fins, and axial musculature simultaneously and cooperatively for swimming like other fish. On land, they only use pectoral fins.

Physiologically, mudskippers must also adapt to living in the intertidal zone known to contain high concentrations of ammonia. They are remarkably tolerant fish with respect to exposure to environmental ammonia. To overcome this potentially harmful environment, mudskippers possess various strategies to ameliorate ammonia toxicity. Their body has saving mechanisms to increase ammonia excretion in the form of urea, hence maintaining the concentration of ammonia in the body at a safe level.

Another mode of movement equally as fascinating is by the starfish. Also known as sea stars, they are one of the most beautiful looking creatures often encountered in the intertidal zones as well as in deep waters. Normally, they are found free-living at the bottom of the sea amongst the coral reef communities. All starfish are carnivores, feeding on other smaller animals, including small crustaceans and molluscs. During ebbing tides, they are occasionally found on the sandy or muddy bottom of the mangroves crawling over rocks, coral rubbles, or shells. There are over two thousand different species of starfish exhibiting a variety of colours, shapes, and sizes. They are not fish as their nickname starfish suggests. It would be more appropriate to call them sea stars instead because of its star shape as viewed from above with five arms, or rays, radiating from a small round body from the centre outwards.

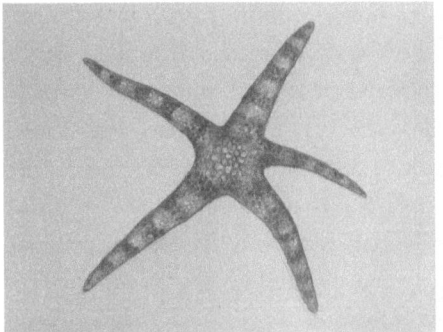

Despite not having blood or a brain, starfish have evolved an unusually sophisticated means of getting about. It has a network of specialised canals filled with water radiating from the centre, which work together to ultimately produce hydraulic pressure capable of lifting its hundreds of 'feet' to initiate movement.

Starfish have a surprisingly unusual sensory anatomy, with no brain or blood. The only sensory organ it has is a set of five purple eyespots strategically positioned at the terminal end of each arm. These eyespots are able to detect the relative intensity and shades of light. Seen almost motionless when the water is ebbing, starfish are not lost or stranded helplessly on land as often thought. There is no cause for alarm if you see one during one of your beachcombing strolls. These creatures may appear lethargic and tired, but rest assured they will make it back home slowly and surely.

What is most impressive about starfish is their mechanics in getting about using hundreds of 'feet' on each arm. I still remember disclosing this fact to a group of children on Lok Kawi beach in Sabah during one of those nature trips by the Malayan Nature Society, Sabah Branch. The children were totally confused trying to make sense about the presence of hundreds of feet on a single arm. Humans have two feet and two arms and are able to run and achieve activities quickly. But here is a creature with hundreds of tiny little 'feet' that only moves at an almost unnoticeable pace. The children thought how totally bizarre nature could be!

I, too, never fail to be utterly captivated by the way nature works. In my opinion, how a starfish executes its movement is one of the ultimate ingenuities of nature. I found myself squatting down on the beach intensely observing a starfish pull itself from the sandy ground. In full admiration, I was recalling what I had read on how starfish moves around. Indeed, starfish locomotion does qualify in the top category of wonderful works of nature. It is a result of its novel vascular system. The incredibly sophisticated design of its vascular system is a work of art. All along its five arms, the system comprises a set of specialised canals that are pumped with water in a way equivalent to our vascular system constantly filled with blood. It has a sieve-like plate called madreporite located on the lower surface, which functions like a mouth. Water from outside is filtered through the madreporite through minute pores and transferred to this network of canals, each having a specific function. Starting with a canal called the stone canal located just beside the plate, water is transferred to a subsequent canal called the ring canal shaped like a pentagon. From the five points of the pentagon, the ring canal spreads out to connect with the radial canals that extend to the end of each foot of the starfish. Most of the task in causing movement of this creature is undertaken by the tube foot. These are the two double rows of 'feet' found along the radial canals. The flow of seawater through this network of the vascular system is crucial for the starfish to move. It relies on the hydraulic pressure created when water is drawn in. The organism is equipped with tube feet called ampulla, which become swollen when filled with water. They taper at the base to form a podium, each having several suckers, which assist the starfish to grip the ground.

The process is elaborate and highly synchronous. The water filtered at the madreporite needs to pass through each of these specialised type of canals before reaching the ampulla. The water is maintained in the ampulla at all times until the organism wants to move. As and when required, the ampulla contracts to force water into the podium till it substantially bulges and elongates. The net effect to all of this is the detachment of the fatter and longer arm from the surface of attachment. Effectively, this frees the arm from its point of anchorage. The creature then brings about displacement from its previous surface contact, seeking a new point of contact. To do this, the ampulla relaxes to form suction cups, which in turn creates a vacuum, allowing it to anchor. This is repeated many times till the starfish is hauled to another location of interest—a novelty in animal locomotion like no other. Note that the movement is manifestly slow, but the mechanics and ingenuity that achieve locomotion are simply magical!

It is not hard to agree with the oft-used axiom 'Have legs, will travel' under normal circumstances and environment. But around and about the mangroves, this might not always be the case. Means of locomotion seen here can be starkly out of the ordinary. The powerful wings of birds of prey, the crutches-like pectoral fins of mudskippers, and the hydraulic power to move hundreds of legs of starfish were a full-on for my tired brain to fathom. If this is not sheer amazing, I don't know what is.

7

Out Of Harm's Way

Passing of genes from the parent animals to their offspring can only occur when they are still alive. Genes can't be inherited from dead animals. Hence, staying alive and out of harm's way has been the preoccupation of all animals. Evolution has resulted in both morphological and behavioural means to avoid the risk of being eaten. The perils of predation inflict strong selective pressures on all species in the wild. To perish and disappear in their environment is to be avoided. Species must take all means and precautions to avoid becoming someone's dinner and thus contribute to the continuance of their own species. The young ones are especially vulnerable. Birds, for instance, build warm and comfortable nests to take care of their eggs and young ones early in life. Hornbills build nests in the tree cavities of tall trees. The female is sealed inside the dark hollow nest behind a mud wall with just a narrow slit wide enough to pass food for the mother and hatchlings. This way, they are kept safe, away from predators that feed on eggs and young birds such as monkeys, raptors, snakes, and lizards. Both parents care for the young ones until they are ready to fly. The obsession with providing safety for the young ones is not limited to mammals and birds. The poison dart frogs in the Amazon forests piggyback their tadpoles to water.

Generally, terrestrial animals are better mothers by way of providing care and safety to their offspring. Caring and concerns for their vulnerable young ones are understandable considering terrestrial animals produce few offspring in each generation. Primates give birth to one infant each year. The orangutan mother, for instance, would stay with their young for up to eight years, teaching them where to find food and how to avoid predators. This is in contrast to the marine animals. Fish and other marine life simply cannot afford the time and labour to care for their young. Instead, marine fish and crustaceans produce vast numbers of eggs and leave the egg's survival to chance. Each of the hundreds or thousands of hatched eggs would further undergo variably complex phases of growth and development. The larval and young juveniles are left to fend for themselves to learn the survival skills of finding food and avoiding predators. But that doesn't mean the parents make no effort to provide the best homes for the development of their young ones. Marine animals use a safer and more hospitable environment to lay their eggs and to allow their juveniles to develop. The mangrove forests have served them well in this respect.

The district of Pitas in Sabah is recognised as the poorest area in Malaysia with many rural communities still living within the hard-core poverty category. Numerous attempts have been made to uplift the livelihoods of the primarily rubber producer smallholders there. Aimed at improving their general well-being, many government-sanctioned development projects have been initiated. But in 2013, the local communities were up in arms protesting about the mangrove clearance for shrimp farming. The conflict was with the local communities that depend on the mangroves for their livelihoods. They demanded the government officially recognise their rights to the remaining mangroves and prevent further clearance for aquaculture. I have had a long interest in understanding similar conflicts involving the socio-economic development versus the natural environment, particularly the mangroves. So I decided to get a first-hand assessment of the situation.

Upon my arrival at the mangroves in contention, I immediately recognised the paramount importance of preserving the ecosystem. For generations, the people here have relied on the fish, crabs, shrimps, and other seafood for their own consumption as well as commercial exploitation. The mangroves here have been the safe home grounds for these marine species. Mangroves here have served as the nursery where adult marine life from the open sea came to lay their eggs. The ecosystem provided their hatched larvae and juveniles with not only sufficient food for growth but also safety from predators—indeed a crucial environment to ensure the safety of the next generation and continuance of the marine species, which the local population depended on. The people wanted the clearance and draining of the mangroves for conversion to hundreds of aquaculture ponds ceased.

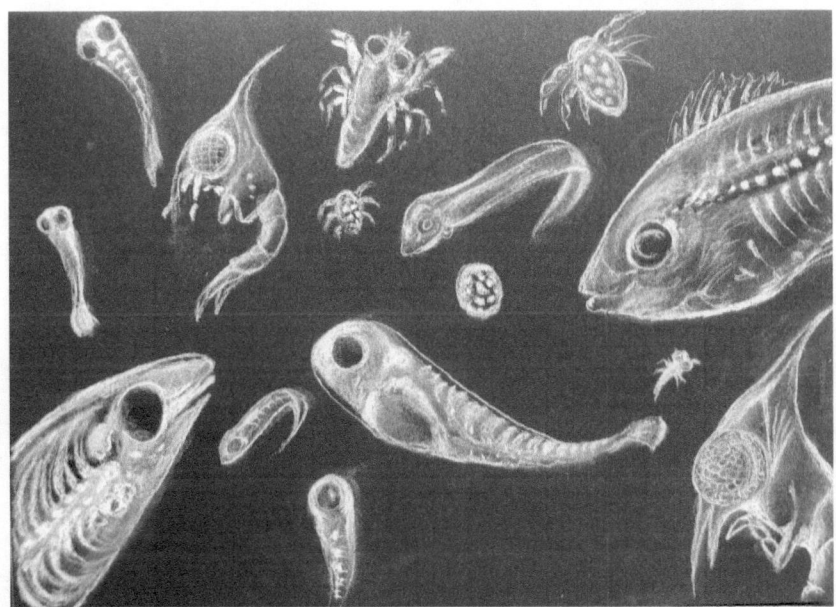

Mangroves are essentially a nursery ground for a variety of marine life to lay eggs and allow the early stages of growth for their species to thrive to adulthood. The ecosystem provides an abundance of food for the hatched larvae and juveniles safe from predators found in the open sea.

The early stages of growth for most species of marine life rely primarily on the abundance of decaying plant and animal matter from the land. The larvae and juveniles of marine animals are slow moving. In the open sea, they are incapable of travelling long distances quickly and widely in search of food. Adult marine species find the mangrove environment most suitable to lay their eggs and leave them to be hatched. The complex root systems of the mangroves slow down the flow of tidal water from the sea to almost stagnation. The calm brackish environment creates a perfect nursery for the larval stage of many species to thrive and remain protected from possible predators. Juvenile fish, shrimps, and crabs feed on dead plants and animals that have fallen to the bottom layer of the mangroves. Here, the larvae turn scavengers, making their meal from floating debris consisting of dead plant and animal materials. Most larval stages here are tiny shrimp-like creatures that feed by scraping food off the rocks and sandy substratum. They have their chainsaw-like tongue to masticate the tough plant debris into ingestible forms. Such diet makes sense because the growing larvae do not need to spend energy hunting but simply feast at will until they are full and well nourished.

At high tide, adult marine creatures are also found in the mangroves to feed off the abundance of plant debris in this ecosystem. Starfish seek to prey on shellfish and small crabs before returning to the open sea with the outgoing tide. When the tide recedes, the beach will be littered with seaweed and dead animals that have been washed ashore. Now it is the turn of the beachcombers to make their presence. Ghost crabs can be seen scurrying over the sand and mud in search of food. In groups, they were seen feasting on dead carcasses of animals washed ashore during the high tide.

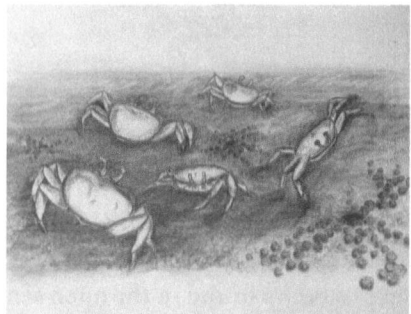 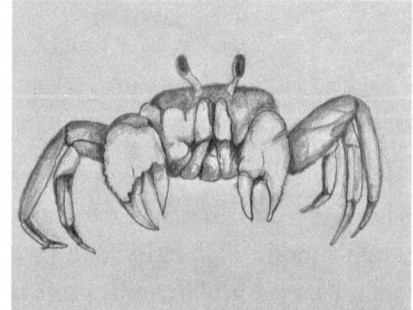

The ghost crabs are generally nocturnal, although sometimes they make daytime appearances as well. They are generally pale in colour and interestingly have the ability to change colour to blend in with their surroundings.

Surviving on vast open spaces like the shore and mudflats of the mangroves can be one of the most challenging feats for small creatures. They can readily become prey to predators because they are more conspicuous against the extensive monotonously coloured landscape. Avoiding becoming prey to the roving shorebirds constantly prodding and picking at anything edible can be a constant contest. Small creatures like fish, crabs, lizards, and frogs can be easily spotted from above by the ever-vigilant birds of prey, which might suddenly swoop from nowhere making a meal of them. For prey species living in an exposed environment, the best defence is to have a hard impenetrable outer layer to protect themselves from both the searing hot sun and predators. Crabs have shells that function as body armour. Starfish wear a leathery skin made from calcium carbonate, a strong tough material for protection. Some species of starfish even go a step further to don numerous spines on their skin to deter predators. Potential predators often find the exoskeleton shield a formidable deterrent, problematical and cumbersome to break before savouring the juicy meat underneath. The attacker often gives up and moves on to an easier target.

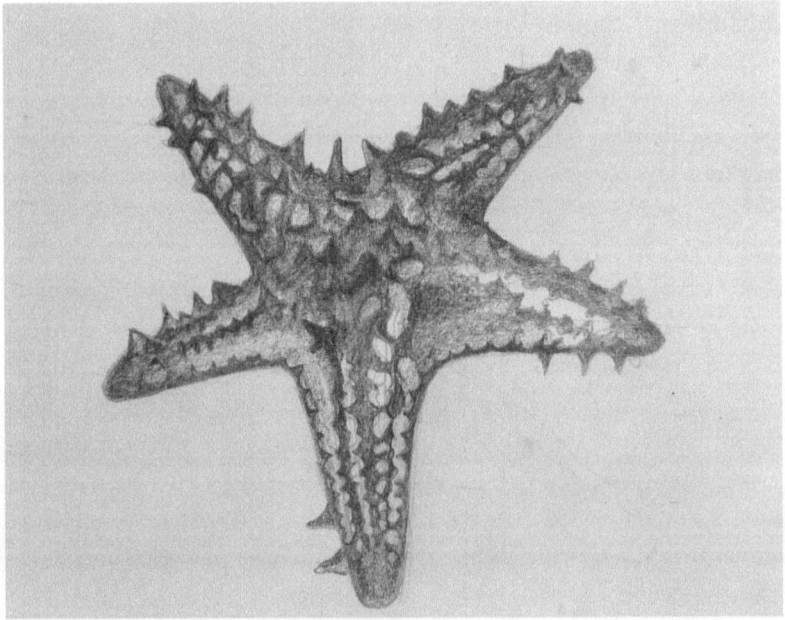

A few species of starfish don a leathery skin with numerous spines along their five radiating arms. Potential predators often find such exoskeletal shield a formidable deterrent, problematical and cumbersome to break before savouring the juicy meat underneath.

A few steps forward, I was treated to another bewilderment of life on the shore adjoining the mangroves. As if inviting me to indulge in a pastime of hide-and-seek, dozens of hermit crabs were darting in and out of their shells as I approached. Slowly taking a squatting posture, I peered at a relatively large hermit crab sitting by the edge of a puddle. It seemed dubious about my presence. I just needed to confirm the fascinating act of drinking by this creature as described in a book I once read. Exactly as I had expected, it dipped its claw in the water and lifted out drops of water to shower its gills and mouth—such a deliberate and animated act of drinking. That, I thought, was truly an awesome sight from primitive little creatures whose nearest relatives are spiders and lobsters. There are over six hundred species of hermit crabs known to science. They are crustaceans, which means they are related to crabs, lobsters, and shrimps. Despite its name, the hermit crab is not a true crab. Stripped of its shell, a hermit crab looks very similar to a lobster rather than a crab. They are known to live for up to fifteen years, although the majority of them die within a year in captivity.

In 1988, I was in a team of researchers carrying out an environmental impact assessment (EIA) study to recommend mitigation measures to the Sabah state government if Pulau Sipadan, an idyllic island off the east coast of Sabah, were to be promoted an international scuba diving destination. There, I saw for the first time the largest hermit crab in the world, the coconut crab, *Birgus latro*. It was about forty centimetres long weighing close to four kilograms. I observed first-hand the sheer power locked in the two huge claws of a coconut crab. We decided to hold it confined in a large plastic pail overnight. Unfortunately, we woke up to an empty container with the captive escaped, nowhere to be found. While we were asleep, our rare specimen managed to push up and topple down two reasonably heavy granite boulders placed over the lid of the pail before we had a chance of documenting the capture in our record book. The powerful coconut crab heroically escaped before giving us its vital statistics!

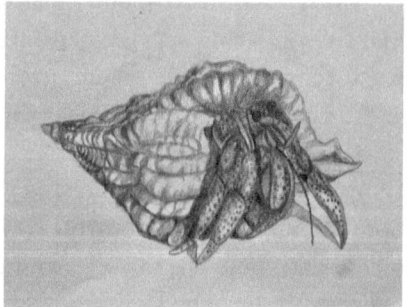

Hermit crabs belong to the crustacean group related to crabs, lobsters, and shrimps. Despite its name, the hermit crab is not a true crab. Stripped of its shell, a hermit crab looks very similar to a lobster rather than a crab.

I silently empathise with the plight of the hermit crab as a living creature trying to survive. Think about this scenario. You have to look for a new home every time you find the present home you're living in too small. Space is not a luxury anymore. You simply cannot function comfortably in your own space at home. Perhaps your family has grown or your household belongings have taken up all the space available. That's not all. You find out that the neighbourhood is notoriously unsafe. There's a high chance you will get killed if you were to be seen outside your home. But you desperately need to be move about as well because you need to obtain food and fresh air. Living at the same address where you cannot freely get enough food to survive is causing concern. So what option is open to you now? Ah yes, a mobile house like a motorhome, which you can drive around in. You can eat, sleep, and undertake your daily chores without venturing out of your home. When your new motorhome gets cluttered and too small for you again in a few years, you will just get a bigger motorhome. Great solution to an ever-occurring problem!

The scenario above is exactly the predicament faced by a hermit crab. They are fascinating creatures as they go through life periodically swapping homes as they grow. They are compelled to abandon the shell they lived in previously and search for a bigger shell for a better fit. The shape of the next chosen shell would be the one capable of snugly fitting the shape of its abdomen. Bizarrely, they find shells that do not even belong to the same class of animals as they do. Because they do not make their own shells, hermit crabs make homes and seek shelter in the shells of another group of marine animals, the gastropods. What a cheek! The shell hermit crabs carry around on their backs are from gastropods, which include snails, slugs, limpets, and sea hares.

There are terrestrial and aquatic hermit crabs, which breathe using gills. Aquatic hermit crabs obtain their oxygen from the seawater, while terrestrial ones need a humid environment and keep their gills moist at all times to obtain enough dissolved oxygen. Some hermit crabs found crawling on the beach can still be a marine hermit crab. Like other crustaceans, hermit crabs undergo a moulting process to grow. This involves shedding their old exoskeleton and start growing a fresh one. But this new exoskeleton takes time to harden. The hermit crab is the only crab that does not develop a hard external shell on its abdomen like other crabs. It therefore needs to be protected from predators. During this soft-bodied phase of growth, the hermit crab has to bury itself in the sand, hidden and safe from ending up on the menu of natural predators, which include fish, squid, octopuses, and shorebirds. At the stage when it is ready to find more food in the open environment, it needs to have a tough armour to protect itself from predators. This comes in the form of a hardy exoskeleton of the gastropod. But honour and dignity still prevail. It would never steal an already occupied shell. Instead, it is only interested in a suitable vacant shell. Upon encounter with an empty shell, it will first assess whether the find fits ideally for its body size. It will closely check it out with its antennae and claws. Satisfied, it will hurriedly grab the shell and slide its abdomen into it. From then on, it will continue growing as it crawls around in search for food.

Everywhere it goes, it will carry with it the tough shell of a dead gastropod on its back. When threatened, it withdraws into this borrowed shell, leaving the predator to try to break the shell. In vain, the frustrated attacker will leave it alone. The hermit crab will only emerge from the safety of its borrowed shell when the predator has departed. It will always be on the lookout for new shells to accommodate its growing body. Each time it swaps shells, the motorhome of choice gets bigger, more spacious, and safer as it goes through life in a hostile neighbourhood. It is cumbersome but an absolute necessity. What a life!

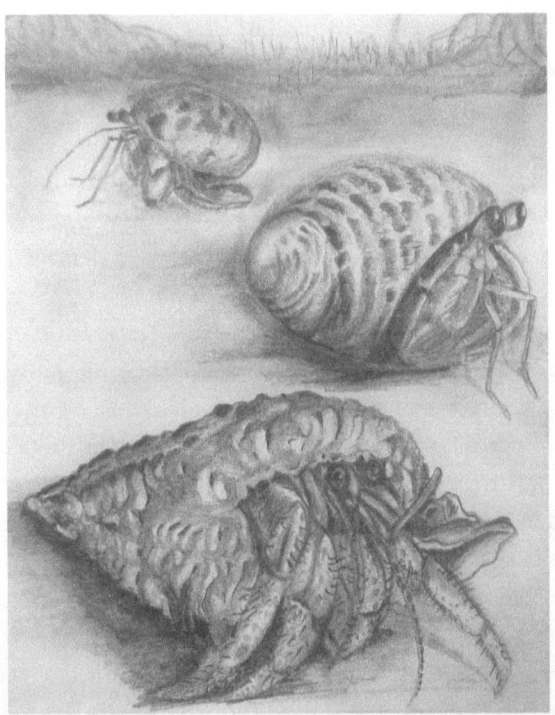

Hermit crabs line up from the largest to the smallest to exchange shells. The largest crab enters the new home that fits, leaving its empty shell behind as the next smaller crab assesses the size of the vacated shell with its pair of antennae and appendages. It will only occupy the shell if deemed large enough for its soft body to continue growing till the next shell is found. The shell swapping continues down the line to upgrade and crawl away with one that fits.

It was past midday. The heat and humidity were strenuously draining my body fluid, and I felt dehydrated and exhausted. Rubbing off sweat from my throat, I felt a sudden urge to quench my thirst and reach for my water bottle from my backpack. I leaned back against a small branch but had clumsily brushed my arm against some leaves. Out of the blue, I jumped in a hasty retreat as I sensed numerous razor-sharp stings underneath my chin. To my big surprise, I had just inadvertently committed an act of aggression, and a war had been declared against me. 'I swore it wasn't a pre-emptive strike!' I felt like yelling out. Regardless, I was fiercely attacked, ripped, and bitten by dozens of ferocious weaver ants that seemed to have emerged from everywhere. They seemed intent as they angrily tapped their feet on the leaves. I wasn't counting, but I'd swear there were hundreds of pillaging weaver ants marching hurriedly out of green balls of leaves, each unwavering in sending a serious warning for me to get off and stop leaning against their nests. Sweeping them off my neck and arms, I obligingly heeded the admonition, took a few steps back, and folded my arms in submission. *My salute and utmost admiration for these little creatures for their strength and ferocity in defending their homes*; I ceded defeat.

But wait, there was another aspect that made these little creatures truly earn my deepest respect. If there was any truth in the maxim 'Every individual has a place to fill in the world', it had to be the world of the weaver ants! Weaver ants collaboratively sew together leaves to build nests in trees. The way they achieve this is sheer ingenuity. A row of worker ants will haul two or three leaves together with their mandibles and feet before sealing the edges to form an almost spherical hollow nest. It may seem cruel and abusive, but it amazed me discovering how they sew the leaves together. They exploited their own larvae to do it. Another row of worker ants, each holding a live ant larva in their jaws, would squeeze out strands of silk from the larval salivary glands and with that silk proceed to stitch the nest together. Without wasting time, these creatures really put their hearts and souls into making a home.

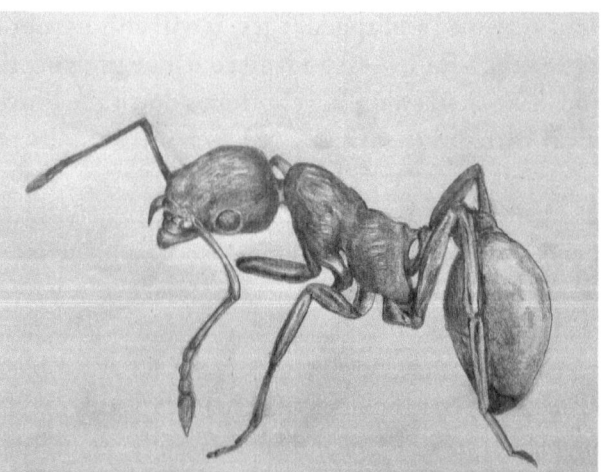

Weaver ants are extremely territorial and aggressive towards other insects. Farmers across Asia and Africa have put this intrinsic trait to good use as a means of protecting their harvests from insect pests like

fruit flies. Innovative ways have been tried to encourage weaver ants to build nests across the entire orchard, since they will only reliably attack when feeling their nests are threatened.

Mangroves have a number of ant species. There exists a type of plant that forms a symbiotic relationship with ant, the *Myrmecodia*, commonly known as ant plant. The *Myrmecodia* plants are usually found bulging out from trunks or branches in the tropical rainforest. But because the branches and trunks of mangrove trees are generally lower than those of the inland forests, I have seen *Myrmecodia* more frequently in the mangrove forests. The plant is easily recognisable from a distance because of its massive bulbous base that protrudes conspicuously from the host plant. Growing in such a position above ground, the ant plant does not have any visible roots like those growing on the ground. But they still need mineral nutrients for growth. It accomplishes this by playing host to ants. The thick globular structure provides lavish accommodation for ant lodgers, and in return, the guest ants living inside provide the much-needed growth nutrients for the plant. On the outside, the swollen globe dons sharp prickles, which deter predators in pursuit of ants for their meals. The surface is also dotted with tiny openings, which serve as gateways through which ants enter or exit their homes. At most times on a hot sunny day, ants swarm all over the globe scurrying in and out of these holes. Most interestingly, inside this home is a network of interconnected chambers, which serve as the ants' living quarters. The queen ant would be sitting calmly in one of these chambers producing her eggs. Essentially, the many chambers found inside this globular structure act as nurseries where the young larvae are kept and reared.

Dissecting the globular *Myrmecodia* base reveals the specifics in the ant plant symbiotic relationship. There are chambers having smooth light-coloured walls and others with darker walls covered with short warty projections. Only inside the dark-walled chambers will the ants deposit the remains of their insect meals and their droppings. Both are rich in phosphates and nitrates. The plant is able to absorb these nutrients through the walls of these chambers for growth. This constitutes *Myrmecodia*'s amazing ability to grow and flourish even in their impoverished locations on tree trunks and branches totally removed from the nutrient-rich soil below.

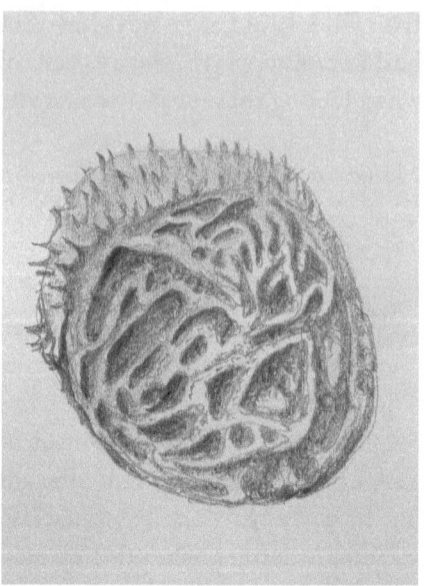

The mangrove ant plant, *Myrmecodia*, is able to grow on branches and trunks removed from the ground because of its symbiotic relationship with ants. The ants deposit their faecal droppings rich in phosphates and nitrates for the plant to absorb as growth nutrients through the walls of a network of chambers found in the bulbous base of the plant.

I continued on my exploration. The air was getting incredibly saturated with humidity. As I was lamenting to myself how the hot humid mangrove was getting to wear me down, I suddenly stopped on my tracks noticing a black object hanging from a branch about fifty metres ahead. Against the pale-green mangrove backdrop, it could be a big animal. It was a semi-circular mass of about a one and a half metres across. But it was not moving and seemed to just hang still off the branch. I reached for my binoculars to confirm my hunch. It was quite an anticlimax when I figured it wasn't an animal of any sort but a huge beehive. I cautiously approached to investigate.

Since moving to New Zealand, I have become quite curious how honey making has become a thriving industry, especially the famous manuka honey. This ultra-sweet golden syrup produced by honeybees has been consumed by humans since the beginning of civilisation. Honey has always been held in high regard. The Bible refers to heaven as the 'Land of Milk and Honey'. In ancient times, it was revered as the food of the gods fit for offerings in sacrifices. In the Roman era, it became a symbol of wealth and happiness. An alcoholic drink called mead was brewed from honey affordable only for the noble-born aristocrats. Honey's antiseptic properties found their use in covering wounds inflicted during wars before the arrival of sterile bandages. Chemically, honey is made up of approximately 20 per cent water, 80 per cent glucose and fructose, pollen, wax, and mineral salts. Its composition and colour vary with the variety of flowers from which the bees have been collecting the nectar. What impressed me most about honey making is to learn how hard the bees have laboured for us. For a single bee to give us a teaspoon of honey, she must have gone buzzing around at least five hundred flowers and filled her 'own stomach' more than sixty times. Estimates are for a hive to produce a kilogram of honey, the hard-working bees, as a team, will have flown the equivalent of three times around the globe. That is incredibly hard work.

I love the bees. They are amazing builders of homes, intricately complex and precisely constructed. Bees build their own homes to protect themselves from the weather and predators. Males usually go into laborious feats to construct convoluted yet purposive structures to impress the females. Honeybees particularly live and work together in large groups in homes called beehives. Just one female, the queen, lays the eggs, and all the other bees are her children. The entire population, except the queen, comprise hundreds of worker bees. They communicate and cooperate to build one huge nest for the queen. The worker bees are great at multitasking as they tirelessly maintain the honeycomb, find food, and raise the young.

Bees abide a systematic regiment in feeding their next generation. Developing larvae are initially fed with royal jelly produced by the glands of worker bees and sequentially with pollen collected from the surrounding flowers. Finally, the growing larvae are fed with honey made from nectar, which has been stowed away in special cells of the honeycomb nicely capped with wax.

The easiest job goes to the queen, which lays a single egg in each hexagonal cell of the honeycomb, which is almost immediately cared for by the worker bees. The next generation of bees is being systematically fed and cared by the present generation in the most consistent and time-tested fashion. Developing larvae are first fed with royal jelly, a special concoction produced by the glands of worker bees. Sequentially, they are fed with pollen collected from the surrounding flowers before graduating to feeders of honey made from nectar. Prior to use as food for the larvae, honey is stored in cells of the honeycomb nicely capped with wax. Apparently, bees have a method to make their hard work slightly easier. The honeycomb has special arrangements for the hexagonal cells. The bottom portion of the beehive would be the broad chambers for the eggs and larvae. Further up would be where nectar is stored, and the topmost part of the hive is last of the menu, the nutritious honey. But there is always a limit to what a honeycomb can hold. When the hive becomes too crowded, scouts are sent out from the swarm to find another suitable location for a hive. The queen then leaves with thousands of workers to create a new home. *Humans may be a social animal, but these social bees represent an absolute epitome how we humans should live;* I nodded in deep thought.

Not near the sophistication of honeybees, I could only think of another type of insect that will work as hard in building homes. The wasps, too, laboriously spend as much time as bees to build homes comprising hundreds of hexagonal cells neatly arranged to form a big structure. The hive is much smaller than a beehive, and the building materials differ in that wasps make their home with sticky spit. They mix their saliva with wood fibres to form paper-like hexagonal walls of their nest. Into each of these cells, a single egg is deposited, and when hatched, the young wasps are raised within this confinement. It is the responsibility of the adult wasp to feed them with partially digested caterpillars.

As pollinators, wasps are important ecologically for humanity. Besides pollinating flowers and food crops, wasps play a crucial role in global food production by keeping populations of crop pests at bay as they feed on caterpillars and other insect pests.

Wasps are often confused with bees because they have a similar shape. The distinction is in the bands around their abdomen. Respectively, wasps have clear yellow and black bands, whereas bees possess less distinct brown and yellow bands around the abdomen. In addition, bees have hairs over their bodies, which facilitate pollen sticking and therefore being gathered more easily, whereas wasps generally tend to be hairless and shiny in appearance.

Animal homes are found everywhere possible in the mangroves, if one cares to search. The estuaries and muddy banks of the mangroves are ideal for the economically important mud crab, *Scylla serrata*. They live in burrows most of the day because they must stay moist or else most will succumb under the punishingly dehydrating hot sun. At night, they crawl out of their daytime seclusion to feed mainly on fallen leaves, fruit, flowers, and seedlings of the mangroves. They are avid scavengers. Hence, they are attracted to any carcass of land or marine creatures in the environment. One moonlit night on Pulau Tiga, Sabah, I stumbled on a few dozen mangrove crabs feasting on the carcass of an adult sea turtle washed ashore with the tide.

Mud crabs displace sediments and mix particles when they create their burrows. Ecologically, they are invaluable in playing this role by engineering the physical and chemical properties of the mangroves. They help to increase the surface area of mangrove sediments, providing more oxygen for bacterial communities to break down organic material and recycle nutrients.

Not exactly visible with the naked eye, there are homes even on the surface of the water. These are floating nests. During breeding seasons, resident waterbirds build their nests along the edges of the wet portions of the mangrove forests. Siting their homes on waterlogged areas serves a good strategy to keep their eggs and chicks safe from land predators, which usually roam on dry land. The nests are made of floating dead vegetation and living water plants. On many occasions, I have tried looking for nests of waders such as white-breasted waterhen, *Amaurornis phoenicurus*, and common sandpiper, *Actitis hypoleucos*, on my trips to the mangrove areas. I have seen adult waterhens darting in and out of the grassy undergrowth but have never seen their nests. They must have cleverly built the nests concealed, not in plain sight of birds of prey or 'predatory' humans.

Tens of millions of wondrous migrating birds use the wetlands to transit on their flight heading for warmer weather and breeding grounds. Some travel relatively short distances, while others cover thousands of kilometres. These wetlands including mangroves, mudflats, lakes, lagoons, and rivers are under increasing threat of drying out and of pollution, leaving migratory birds at growing risks.

Our most unassuming pesky mosquitoes also build homes that float. All mosquitoes need water to complete their life cycle. It would be an evolutionary advantage for mosquitoes to start life amidst a wet watery environment. As part of their life cycle, mangrove mosquitoes build a floating raft comprising more than 100 eggs on stagnant water and pools found throughout the mangrove forest. They lay their eggs one at a time on the water surface and skilfully bind them together in a floating raft strong enough to withstand the mosquito's weight.

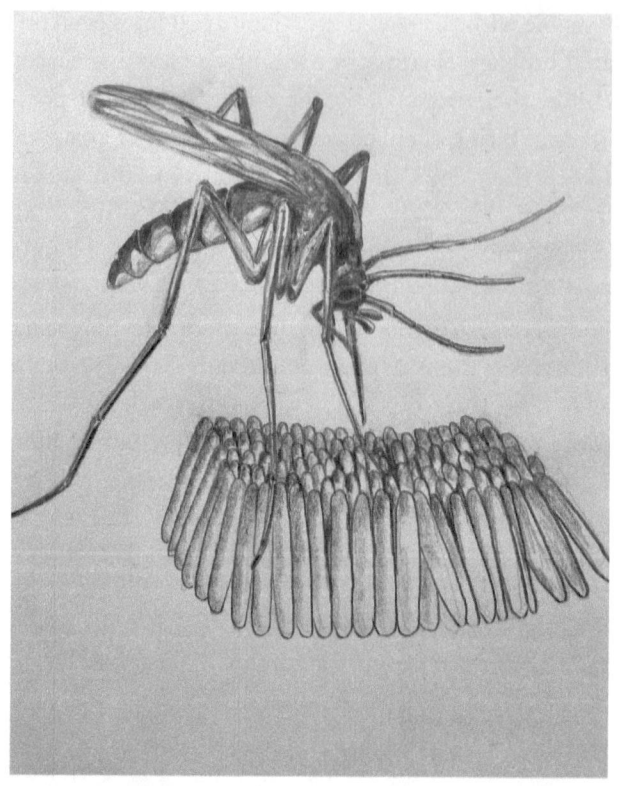

Mosquitoes need water to complete their life cycle. Mangrove mosquitoes build a floating raft comprising more than 100 eggs on stagnant water and pools found throughout the mangrove forest.

The wet undergrowth of the mangroves is also idyllic for crocodiles to build homes for their young ones. This huge beast will pile soil or sand mixed with plant materials to build their nests. The mound acts as an incubator for their eggs. During the day, the nest retains heat from the sun, and during the cooler rainy season, heat is generated from the decaying plant matter within the nest. This is important because the sex of crocodile hatchlings is determined by the temperature during egg incubation. Slight changes in the temperature will result in all-male or all-female hatchlings. Crocodiles are also amongst the world's most devoted mothers in the animal kingdom. For almost four months, female crocodiles will fiercely guard their nest from any intrusions. Newly born hatchlings emit bird-like chirps to indicate they are ready to be dug out.

From the tiniest mosquitoes to the very large predator crocodiles, the mangroves are home to many creatures. Marine life far out at sea relies on the mangroves as a safe haven for their eggs, larvae, and juveniles to fully develop into adult forms. This unique biological function is often ignored by humans in pursuit of socio-economic development. A compelling question begs an all-encompassing answer: Can we not prudently and sustainably manage the health and existence of mangroves to benefit humanity in the face of future calamities such as biodiversity loss and climate change?

8

Got You Covered

In recent years, questions have been raised as to whether animals, other than humans, have consciousness. If animals are conscious, do they experience different levels of consciousness? Emerging consensus appears to support humans are not alone; other animals do have some form of consciousness. There are other conscious beings including all mammals, birds, reptiles, and amphibians. Remarkably, sea animals like the octopus, squid, and cuttlefish have been shown to be conscious in that they are self-aware creatures. They are also highly intelligent and capable of anticipating a painful, difficult, or stressful situation. In addition, they are also able to remember those painful experiences.

I found roaming about in the mangroves always rekindles thoughtful wonderment on how evolution has gifted us with those senses. Myriads of clones have been time-tested over billions of years for the betterment of animal species to become conscious of their environment. For wildlife species living in the mangroves, making sense and recognising the mixture of noises emanating from an assortment of other animals within the same environment could present a huge challenge. The effects of these extraneous noises could adversely influence their ability to sense the presence of identifiable food sources and predators. In the long term, that might impact adversely on the species abundance and survival. Thanks to millions of years of evolution and adaptation, the mangrove species have been gifted with amazing senses to maximise their foraging behaviour, increase breeding success, and escape from predators. Various ingenious mechanisms have helped them to become conscious of their surroundings. They don't easily become victims to predators. For food, they are able to listen and locate their prey. They also need not struggle in communicating with their own species to find suitable mates.

I looked down along the coast unrestrained to unwind my eye focus.

Straight ahead along the water's edge, I noticed a long trail of claw marks in the soft grey mud leading towards several stands of *Pandanus* vegetation. Just as I tried to figure out the possible owner of those claw markings, a menacing-looking head of a lizard popped out from behind the *Pandanus* prop roots. Within the next few seconds, a metre-long Asian water monitor lizard, *Varanus salvator*, came in full view. A remarkably handsome reptile, it is amongst the largest lizards in the world, known for their ability to survive in almost any habitat. The species can survive in environments that would have spelled doom for any other large carnivores. Monitor lizards have been found living on paltry isolated islands and solitary rocky outcrops far away in the middle of the blue ocean. Mangrove forests naturally offer ideal habitats for monitor lizards. Their success is credited to them being cold-blooded, enabling them to make more efficient use of ingested food. Warm-blooded creatures must burn fuel constantly to keep their body temperature constant. But cold-blooded creatures can get by with less food because they are able to maintain an almost constant body temperature. This makes monitor lizards energetic and more active than other reptiles of similar size.

Clearly, the lizard was on a foraging expedition. The impeccable functioning of its olfactory system was in full gear, flicking its tongue in and out of its mouth to collect particles from the air, water, and ground. It was smelling the air. The tongue has a discriminating role for picking up the smell of molecules in the air, which could be a question of life and death. Unlike humans, their nose isn't the primary organ used for smelling. It is not keen enough. Their flat nose has small nasal cavities around the snout, which are not sufficiently reliable to detect food odours at a distance like most mammals. Instead, snakes and lizards have a tongue that is tailored for the task of smelling and tasting. The tongue sniffs out edible foods by detecting chemical ingredients in the air. It possesses an ingenious ability to analyse aromas using a special apparatus called Jacobson's organ. It is a shallow groove consisting of many sensory cells situated in the roof of its mouth. By flicking the long forked tongue in and out of its mouth, the lizard is effectively exposing this Jacobson's organ to the air. In search for food, this is the means by which they detect chemicals present in the air, water, and ground. It can detect odours of decomposing animals up to four kilometres away.

A Jacobson's organ is an indentation in the roof of the monitor lizard's mouth consisting of many sensory cells. By flicking its forked tongue in and out of its mouth, it exposes this organ to detect chemicals present in the air, water, and ground. This way, it is able to detect odours of decomposing animals up to four kilometres away.

Of all the five archetypical senses, smell is the oldest. This sense evolved when the first land animal emerged five hundred million years ago. It has its origins in the bacteria known to have the rudimentary senses for detecting chemicals in air and in water. Evolutionarily, animals could smell before they could see or hear. In fact, animals respond to chemicals around them before they can feel by touch. The specificity and precision of their ability to pick up the chemicals in the natural environment underpins their success in adapting to the harsh environment of the mangroves.

More importantly, through learned behaviour, snakes and lizards are able to associate smell with items to be approached as food. In a state of hunger, they can evaluate whether the food is edible or worthy of ingestion using odours. If deemed edible from the smell, they will devour it and continue to seek for more till fully satisfied. However, if the food smells bad or repulsive, that will result in rejection by seeking other sources.

The keen sense of smell not only helps the monitor lizards in finding food but also informs them if the coast is clear. Their tongue has the remarkable capacity in assisting them to differentiate and learn that certain odours ought to be avoided as they may be toxic or make them sick. Smell also comes into play in securing a territory that is safe from competitors and predators. Certain enemies emit characteristic odours that can be picked up by their tongue. Surviving in the mangrove environment, the monitor lizards soon recognise the smells that connote risk or danger. They learn and know when to fear and flee. Within their own species, smell can also help choose potential mates that are nearby and ready to copulate.

The impeccable functioning of smelling and tasting by the lizard's tongue is a question of life and death. It is used to sniff out edible foods by detecting chemical ingredients in the air as well as discriminating aromas associated with poisonous creatures.

Mammals generally have bigger noses and nasal cavities than other animals. Their sense of smell can be more prominent and used for longer-term benefits. Orangutans, for instance, learn quickly to associate smells of burning forests with long-standing danger to their species. They are able to recognise the smell of smoke and make better preparation to flee from being caught in the spreading forest fires to avoid suffocation. Signs of risks or threats like the presence of predators within the vicinity can also be duly smelt and quickly responded to. Certain odours emitted by predators lead them to avoid certain areas of the forests. They can also change their feeding behaviour after learning that consumption of food having a certain odour could end up making them sick. The smell of burning embers in the air, airborne or waterborne chemicals will also indicate the habitats that are not conducive to supporting the growth and survival of their species. This can lead to evading and moving away from the source of the smell.

I was watching a group of long-tailed macaques rummaging through a garbage bin in a mangrove reserve park once. They appeared to be scavenging the leftover food discarded earlier by a crowd of picnickers at a nice spot on the beach. Like humans, there are a number of observable antics animals display before ingesting food. For this, a combination of senses is used. For example, the monkeys were utilising both vision and smell to select discarded items before putting them in their mouths. They would first assess whether it was food suitable for eating. Under conditions where the quality of the food is questionable as judged by its unusual colour, the animal would sniff it a few times before either ingesting or rejecting it. Unexpected off-putting flavours would negatively influence their appetite followed by rejection. All of these decision-making behaviours rely heavily on the olfactory system's capacity to learn from experience.

Generally, birds are poor in their sense of smell. In fact, some birds cannot smell at all. However, there are exceptions. Some seabirds including a few species found in the mangroves are very sensitive to smell. When foraging in the murky waters of the mudflat, their tube-like beaks are useful in finding and selecting food through smell. The migratory birds that use the mangroves to transit during their long journeys are known to utilise their keen sense of smell during migration. Their long arduous flights almost invariably take place throughout the night when they cannot see very well. Then their sense of smell comes to the rescue in several ways. In piloting and navigating their journey, migratory birds need to recognise other individuals on the same flight and stay close to them. This is a means of keeping migrating birds within the flock, not straying or becoming lost or left behind. Their relatively large noses and external nostrils on top of their beaks give them a keen sense of smell in finding food floating on the sea even in total darkness. When on feeding expeditions way out at sea, migratory seabirds also use smell to find their way home by following the unique odour of their individual nests.

A majority of birds have external nares or nostrils located at the base of their beak. They are a pair of non-exaggerated holes in the shape of a circle, slit, or oval, which invariably leads to the nasal cavity in the skull and hence connected to the rest of the respiratory system.

Most reptiles can survive and reproduce in interior terrestrial habitats because they require dry land to lay their eggs. But there are snake species that seem to have adapted to living in mangroves despite the ecosystem being incessantly wet and inundated by coastal waters on a daily basis. The shore pit viper, *Cryptelytrops purpureomaculatus*, is a snake sometimes spotted curling motionless on tree branches in the mangroves. Not once in my many trips to Bako National Park, Sarawak, have I missed seeing this highly poisonous species. About a metre long, it can be recognised from its typical broad triangular head, red eyes, and quite a fearsome disposition. But in reality, it is a shy and reticent snake in nature.

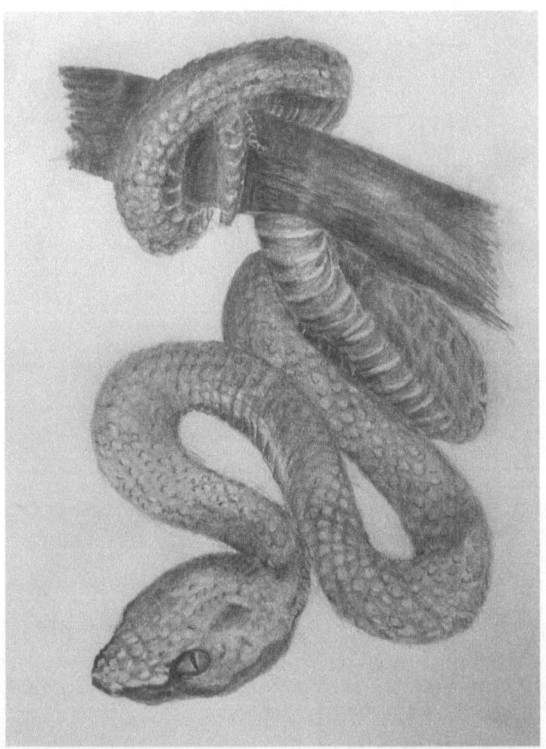

Pit viper 'sees' using a pit organ that can sense infrared thermal radiation. It allows the animal to strike warm prey accurately, even in the absence of light, from several metres away.

The success of this viper species is largely attributed to their ability to sense, seize, and kill their victims quickly. This is aided by being acutely aware of the presence of potential food in their proximity. Its lethargic docile nature can quickly turn aggressive during hunting or when defending itself. It can strike swiftly from quite a distance feeding on lizards, frogs, small rodents, and birds. Like the monitor lizards, the pit viper has a sixth sense, equipped with a Jacobson's organ to detect traces of chemicals that indicate the presence of food or predators. Chemicals are collected on the tip of its forked tongue and delivered to clusters of sensory cells for analysis. The information on the learned nature of the food or predators is sent to the brain in a millisecond for it to respond immediately.

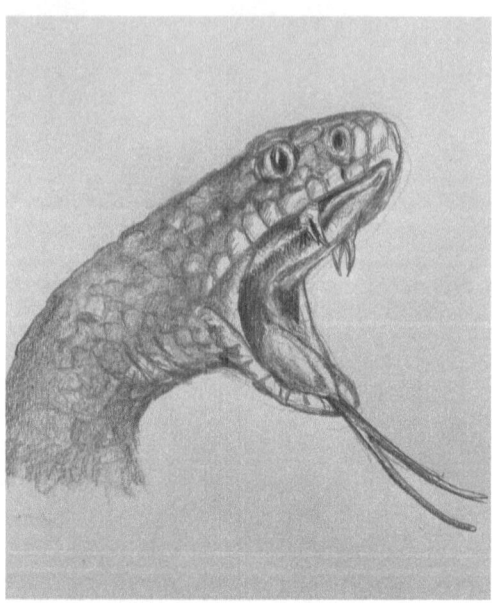

Chemical ingredients detected by the forked tongue are sent to clusters of sensory cells capable of transforming them into information if they are food, dangerous substances, or predators. The speed of these discriminatory messages sent to the brain is so swift that the viper snake is able to respond instantaneously.

As I was silently admiring the many amazing adaptations of mangrove species, I was at the same time constantly persecuted by clouds of mosquitoes hovering over my head, my neck, and my arms. They were vicious and vociferously hungry. While a male mosquito is content to live off the nectar and juices from plants, the female is not. She obtains her nourishment from the blood of her victims. Using her sharp beak, the female mosquito pricks human or animal skin, pokes her proboscis, and sucks in one or two drops of blood as her meal. A hearty blood meal is required before the female mosquito can mature and lay eggs. To maintain a good flow of her blood meal, she will also secrete an anti-blood-clotting chemical during feeding. First-time visitors to the mangroves often suffer unusual itching and swelling after being bitten by mosquitoes. This is more of an allergic reaction to the anti-clotting substance secreted by the mosquitoes than anything to be of serious concern.

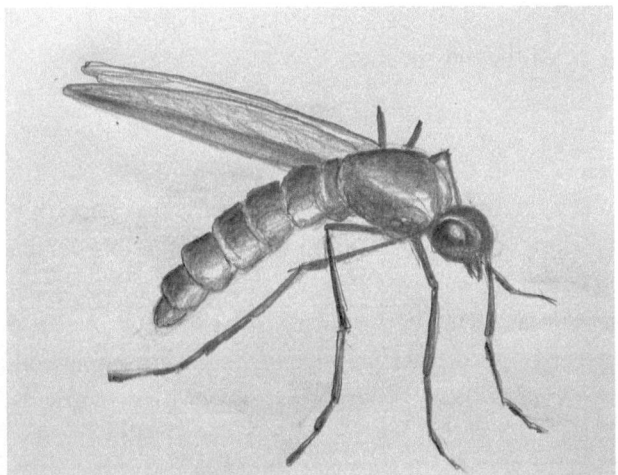

The female mosquito pricks human or animal skin with her sharp beak to have a blood meal, an absolute prerequisite before she can mature and lay eggs. To maintain a continuous flow of the blood, she also secretes an anti-blood-clotting chemical during feeding.

A mangrove swamp in my opinion ranks as the most gruelling to negotiate in the world. As if having to zigzag and move around in the mud was not the most fatiguing thing to do on a humid tropical day, ghastly swarms of mosquitoes making a meal of oneself should be enough to put you off from exploring further—a truly inhospitable environment made worse by harassing bloodsuckers! Why did they seem so intense at biting me but not the person next to me? Exasperated, I violently gave a lightning-speed swipe on the back of my neck. I had to stop assuming my blood was nicer-tasting blood than that of the other people around me. Indeed, mosquitoes tend to be choosy on whom they seek to bite. But the discriminating factor isn't the blood. Mosquitoes rely on both sight and smell to select its blood donor. They use infrared detectors to hit the target assisted by certain chemicals emitted from the skin of the victim. Apparently, they are able to smell and become attracted to carbon dioxide exhaled by potential victims several feet away. Mosquitoes also sense body chemicals, such as the lactic acid in our sweat, and fly directly to make a landing and draw blood. But I came well prepared. A logical strategy to prevent mosquito bites is avoidance. So before leaving home that morning, I had generously smothered my exposed skin with a time-honoured mosquito repellent, which would do just that. It contained the primary active ingredient DEET

(N, N-diethyl-meta-toluamide), which proved beneficial for several hours. That really saved the day for me.

Not all insects are endowed with sensitive olfactory means to help them sense and be aware of their surroundings. An elite few have evolved to see ultraviolet (UV) light, which is invisible to humans. Bees and butterflies are amongst these insects, which also happen to be important pollinators of plants. Flowers go to great lengths to attract pollinating insects. To attract the pollinators, a quarter of all flowers are known to reflect UV light, which can be seen by insects from afar. The centre of yellow flowers reflects the most UV light. When seen by pollinating insects, this UV-reflecting spot acts as a guide leading to their food source—the pollen and the nectar. The petals of the flowers provide convenient landing strips and platforms for the pollinating insects. There is a resounding win-win advantage to this—a mutually beneficial relationship between flowers and pollinators is forged, where the bee is provided with food in the form of nectar or pollen, and in return the plant benefits from dispersion of pollen to other flowers of the same species. This is an example of a co-evolutionary relationship.

Recent evidence suggests bees are in catastrophic decline because of complex issues including indiscriminate use of pesticides, habitat loss, and climate change. Our agriculture and future food security will be seriously threatened with the decline of wild pollinators such as bats and bees.

Other insects function in the world around them by having evolved sensitive feelers such as antennae, whiskers, tentacles, and tongues. Flies do not have long antennae like the beetles, butterflies, or moths. Instead, they have sensory hairs covering every part of their body. Hairs found on their feet and mouth are used to taste. If they happen to land on something edible like sugary foods, they will instinctively lower their mouthparts and take a bite.

Butterflies have evolved such that almost every part of their body is capable of detecting odorous chemicals. They also have taste buds located at the end of the tongue and identify the chemicals using sensory structures on their feet.

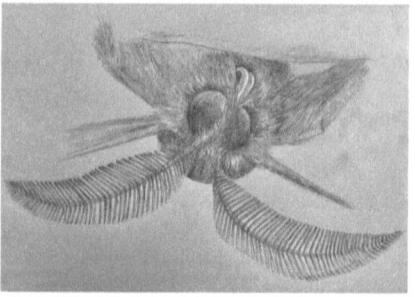

A male moth's antennae are extremely sensitive to detect the female sex pheromone emanating from as far as 4.5 kilometres away. It is also sensitive enough to differentiate female sex pheromone from competing smells of food, predators, and contaminants such as dust.

Sometimes, we fail to realise and appreciate how we humans can perceive vibrations. They are important sensory stimuli that we encounter and detect daily in our lives. We react instantly to the feeling of a mobile phone vibrating in our pockets announcing an incoming call. This is due to specialised receptors capable of converting vibrations into neural signals sent to our brain. Animals in the mangroves also live surrounded by vibrations, which they sense to perceive events in the environment. It is crucial for them to be alert and vigilant at all times. Spiders are extremely sensitive to vibrations. Surpassed only by the cockroaches, spiders are the second most vibration-sensitive organism on the planet. Almost all species of spiders are known to be sensitive to vibratory stimulation emanating from their web. Making sense of these vibrations is extremely important to alert them of pending meals or to the presence of deadly predators.

Spider silk is an incredible building material, which begins as a liquid produced by glands on the abdomens of females. Special organs called spinnerets squeeze it out as fine fibres and twist them into a single thread. The spider's back legs then draws out the silk to construct their web, which has two functions: protecting eggs and capturing prey. Orb spiders can weave their homes almost anywhere. They invariably site their traps where prey tend to be caught in abundance. Many orb spiders are known to accomplish this nightly after recycling their old web by eating it. The silk provides them with the protein as well as the energy they require.

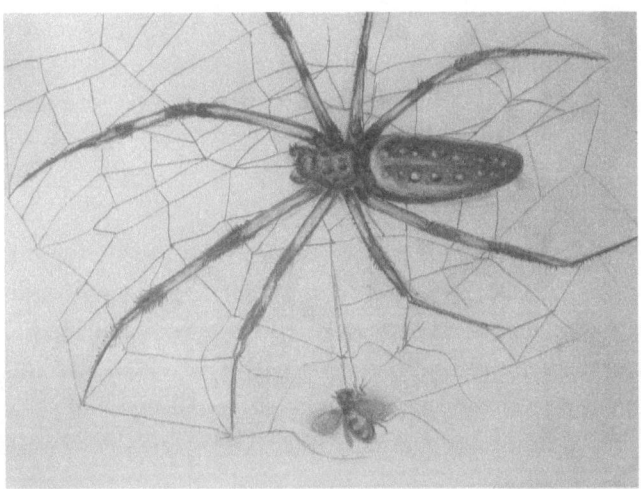

Living on cobwebs in places having low-light penetration, spiders have evolved to become extremely sensitive to vibrations caused by prey. They are the second most vibration-sensitive organism, topped only by cockroaches.

Caught stuck to the web, the struggling prey sends vibrations, which can be felt by the waiting spider along the web. Before deciding to go for a feast, the spider must assess whether the besieged victim would make a wholesome meal with respect to size. If the vibrations are within a defined frequency and amplitude range, the spider attacks the vibration source. But if the vibrations do not fall within the defined strength, the prey would be considered not worthy of response. Under such circumstances, the source from where the vibrations originated would not induce an attack. Once a sizeable prey is caught, the meal is elaborately prepared involving wrapping the victim in silk, followed by injection of saliva to slowly dissolve the captured prey. Besides detecting vibrations, spiders can also see prey even at low-light levels. It is known to detect odours through hair sensors called pedipalps found on their feelers.

Over time, animals living in and around the mangroves have developed winning attributes to improve their becoming alert and vigilant of their surroundings. Why not if this can increase their chances of ensuring enough food in the face of competition with other species or in situations of inclement weather affecting food supply? There are a few strategies observed in nature capable of circumventing these uncertainties. The response can be of short-term or long-term benefit. For sustaining long-term survival, animals may even change their dietary habits like adapting to becoming a generalist from a specialist in their food preference. Finding themselves in a situation of food scarcity is less likely on a generalist diet where shortage of a certain food source can be substituted with another.

Being always on the alert is paramount to the survival of animals in the mangroves with respect to finding food, sensing dangers, and avoiding predators. Alertness has an important role to play in behavioural ecology. Vigilance and covering each other are useful attributes. At all times, an animal's deliberate examination of its surroundings and awareness of predator presence become paramount. Indeed, vigilance is an important behaviour during foraging of the most iconic resident of the mangroves, the proboscis monkey. My visits to the mangroves in Sabah always find me standing with folded arms, admiring how caring and concerned the alpha male proboscis monkey is in regard to the affairs of his group during feeding. There, he perches on the higher branch looking for his several wives and siblings spreading over the mangroves picking young leaves to eat. He seems clearly to be in charge of making sure none on them stray too far from the group. This is the task of the dominant male.

Proboscis monkeys represent an excellent scenario where vigilance is carried out to protect a group of individuals. They forage in groups. A vigilante is charged to keep careful watch for possible danger or difficulties befalling the group. He sends specific alarm signals to alert the group on the presence of predators. He perches attentively on sentry duty, looking for predators, while the rest of the group forage. I can almost hear him saying to his foraging group, 'Go ahead and get full. I've got you covered!' He definitely is a confident vigilante at all times. This requires both physiological and behavioural adaptations. Proboscis monkeys have evolved amazing mechanisms to heighten their awareness of the surroundings in which they live. They do this at the level of group individuals and also of the species.

Living in the wild, most monkeys have a better chance of surviving if they function as a group. Trying to survive without the help of others can lead to many risks. Unsuspecting predators can be lurking from nowhere amidst the lush green backdrop of the mangroves. Group living also aids in finding food. Chances of locating abundant food sources are easier with more pairs of eyes undertaking the task. Proboscis monkeys are amongst the many types of primates known to unite in the wild. It has become a natural phenomenon to warn each other, especially the young, of impending danger or the availability of food.

Infanticide occurs in the proboscis monkey species. An adult male that takes over the harem will kill infants from a male he succeeded. During the transition period, infants are often guarded with extreme suspicion by their mothers.

Mangroves may not be a populous ecosystem akin to terrestrial rainforests, where the density of animals present is high. But the inhabitants of mangroves are diverse in species. Each of these species is constantly generating noises of every description. The surroundings are always pulsating with a myriad of sounds from clandestine creatures. It is a noisy place. Just before the day starts, the dawn chorus of birds fills the air. Combinations of birds chirping and crickets stridulating away make the ecosystem sounds chaotic. The buzzing, whining, and humming in the background do not help when an individual species is trying to listen for the presence of prey around. Animals relying on sounds to detect the presence of food are particularly affected.

Male proboscis monkeys honk as a means of communication amongst the individuals in their group. The honking sound is sufficiently loud to drown the noises in the background and be heard by those of its species. There are a number of variations in his honking to mean different things. It is imperative these honks for special messages be heard and heeded appropriately. But it is relatively difficult to be heard clearly and distinctly amidst the noisy background of the mangroves. Nature has it that his long nose becomes handy in being heard and drawing the attention of individuals in his group. The bigger the nose, the better. The nose acts like a loud speaker or a hailer, which he uses to make himself heard louder in warning members of his troop of any looming threats or danger. His nose becomes enlarged, allowing blood to rush in, filling the nasal tissue, transforming it into a resonating chamber. His females and youngsters in distant locations heed his honks, making his marshalling task more effective. Louder honks from his commanding position high up on the tree would also sound more threatening to other hostile males in the proximity. During the mating season, such lurid and showy honks are more audible for attracting potential mating partners. The males with louder vocalisations are able to communicate with females in distant locations through honking.

Dominant male proboscis monkeys honk to mean different things. A growl is meant to bring peace amongst his squabbling troop members or to issue a threat to other competitor males within the proximity. Another type of honk is specifically issued to warn the young ones about lurking predators nearby.

His honks sound differently for specific purposes. A growl is meant to bring peace to the commotion that might be prevailing amongst his group. To issue a threatening warning to other groups within the proximity, the male proboscis monkey will exact an inhospitable honk. Hearing such aggressive honks, other males would feel intimidated and tend to stay away from the harem. There is yet another type of honking issued to warn of the presence of predators. Loud continuous honking warns the young ones of lurking predators nearby. These may include pythons, clouded leopards, and crocodiles. Not all honks are necessarily loud. Juveniles of both sexes and adult females do not honk but emit a piercing shriek when they become agitated or scared. Discordant ear-piercing screams are also heard during feeding bouts at night just before sleep. In stressful situations like this, the male proboscis monkey is capable of fine-tuning his honks to pacify and assure infants of their safety and well-being. These softer comforting honks are useful in calming his offspring and reducing anxiety within the group. With such over-arching roles of their noses, it is no wonder a long-nosed male proboscis monkey is given much adulation and authority within the group.

For all intents and purposes, vigilance and being conscious of the surroundings are beneficial during foraging and escaping from predators. The proboscis monkeys often need to venture away from the safety of their homes to find food. A useful means of detecting locations of this food will assist. When roaming about, they will always be at the risk of being ambushed by concealed predators. Sensing the presence of their unseen enemies at timely instances prevents them from becoming prey. However, being vigilant comes at the expense of time spent on other things like feeding, child caring, or seeking mates to reproduce. The length of time animals devote to vigilance is compensated by their being endowed with amazing senses capable of detecting subtle clues from the environment indicating food or danger.

9

Staying Connected

Staying connected is everything today. It is even said to be good for one's health and well-being. One bends over backwards to create day-to-day opportunities of meeting others. Nowadays, the internet and mobile phones are used to send emails or text messages to keep current and connected with everyone close to you. We use communication to share information, comment, ask questions, express wants and needs, and develop social relationships and social etiquettes. These seem to be the limits why we stay connected. The need to communicate is universal. It is also normal to all animal species, not only humans. Animals, however, communicate for reasons that are more existential than ours. Communication is very important in the animal kingdom, which is the reason why animal sensory systems used for communication are usually more developed and sophisticated than those of humans. For example, a dog's sense of smell is forty times more acute than that of humans! More importantly, their communication behaviours are inheritable. In other words, their ability to give and respond to communication signals can be passed on to the next generation at birth. This is to ensure that the organism's likelihood of surviving and reproducing will persist in their future clones of species or population. To them, staying connected has to do with their existence. In essence, staying connected is a game of survival. Their meaning of life revolves around the ability to communicate with their own species, ensuring their existence in a more specific way. They need to communicate to find mates, establish dominance, defend their territory, coordinate group behaviour, and care for their offspring.

Communication takes many forms. We know that animals have evolved different senses similar to humans. Like us, they, too, have the classic five senses used to detect what is occurring around them. They see, hear, smell, taste, and touch. We also have a sense of balance and ability to feel heat and pain. Additionally, we have the sense that allows us to perceive the position, movement, and action of our body parts like our limbs and muscles. This sensory mechanism is termed proprioception or kinaesthesia. The development of proprioception in animals is still vaguely understood. Over millions of years, we humans may have developed slight variations in the sense organs to detect these signals. Some animals may have superior mechanisms to us. Their sensory receptors allow them to sense such things as light, temperature, moisture, and movement better than humans. But in general, humans and animals are able to sense signals based on visual, auditory, or sound and chemicals involving pheromones and tactile or touch-based cues.

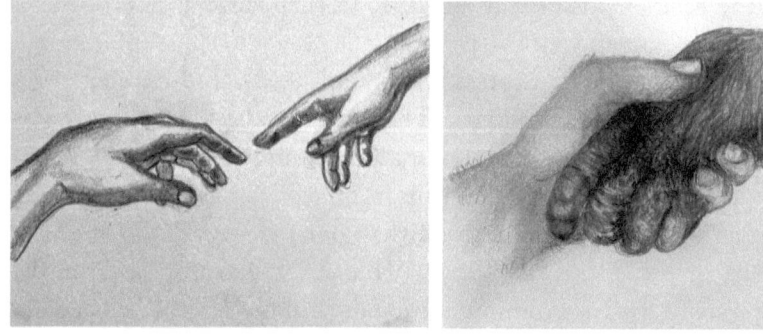

Touch was the first sense that developed in multicellular organisms like animals and humans. Also known as tactile communication, it involves several distinct sensations, which are communicated to the brain through specialised neurons in the skin.

Besides being constantly loud and noisy, the mangrove environment is also smelly. There is an ever presence of a range of volatile chemicals in the air at all times. To us humans and most mammals, these chemicals are olfactory signals that provide the ability to identify and react appropriately to olfactory signals clearly well developed in some of the animals found in the mangroves. The atmosphere reeks of an unpleasant smell of hydrogen sulphide, the same gas responsible for the rotten egg smell. Fortunately, odorous hydrogen sulphide is generally non-toxic. The gas is produced from organic matter breaking down under anoxic conditions. The type of bacteria living in the mangrove soils is primarily anaerobic decomposers— meaning, they are able to perform the decaying process in the absence of oxygen. The by-product of decomposition of plant debris and other organic matter under such conditions produces a variety of odours, primarily the foul-smelling hydrogen sulphide. The cacophony of smells in the mangroves can only be off-putting to humans. But to the animals relying on their olfactory system to communicate, they find this ever-present odour masking their ability to smell. The overpowering mangrove smell can interfere with their effectiveness to communicate and stay connected. The adverse consequences are many, including reduced efficiency in locating food, mating partners, or predators. How have the evolutionary processes helped to overcome this? An alternative solution is to evolve means of communication that do not rely entirely on smell. There is a battery of ways information can be passed from one animal to another. The information must comply with respect to the recipient getting the appropriate signals for specific purposes. They must be recognised and deciphered accurately between members of a species, not by a different species, which may be a competitor or a predator.

Auditory communication based on sound is widely used in terrestrial animals. Birds are one of the earlier groups of animals to use sounds to convey warnings, attract mates, defend territories, and coordinate group behaviours. They chirp and sing to stay connected. The communicating signals birds use are relatively elaborate in the form of a melodic vocalisation and are very characteristic of a species. They sing to attract mates in the breeding season. The males sing their unique melody, which can be recognised by females of their own species. It is important that the arrangement of the high and low notes and the pauses between each segment of the song are in correct sequences. Sometimes, these song arrangements not only are in a general sense characteristic to the same species but can also be distinct to the same species found in certain regions of the land mass. For instance, the same bird species found in the land region of Borneo will sing slightly differently from those in Peninsular Malaysia. This is equivalent to people from different regions of Malaysia communicating in different dialects of the Malay language in the thirteen states of the country. The young male birds are found to learn the unique songs of their species by copying that of their fathers.

Birds are connected using songs and chirps. They are one of the earlier groups of animals to use auditory communication in the form of melodic vocalisation in conveying warnings, attracting mates, defending territories, and coordinating group behaviours.

Animals need to have vocal cords to call or make noises through their mouths. They can vary the pitch and loudness of their calls by varying the opening and closure of the vocal folds found in the windpipe. When they breathe out, the vocal cords open and let the air out of the lungs, causing vibrations, which produce sounds. This is most obvious in frogs and toads, which have a vocal sac, which works like an inflatable amplifier. A frog starts to call by breathing in and at the same time closing its nostrils. This way, it forces the air backwards and forwards between its lungs and vocal sac, emitting a unique call recognisable only by potential mates of the same species. These sounds are used not only for courtship but also for defensive purposes.

Frogs possess vocal sacs, which function like an inflatable amplifier through which they breathe out air from the lungs to cause vibrations and produce sounds. It varies the pitch and loudness of its calls recognisable by its own species by opening and closing its nostrils.

Members of the spiny lobsters, for example, produce sound as a defence mechanism. They do this by rubbing two body parts together, especially the skeletal parts called plectra found at the base of the antennae. To produce sound, the lobster pulls the plectra over the file, which is the rough surface just beneath the antennae. While doing so, it goes into a defensive posturing position accompanied by clicking sounds to scare the predator away.

But nothing beats the sound made by the mangrove snapping shrimp. Ever since I was first introduced to the mangrove environment, I have always been mystified by the loud pop sounds often heard emanating from its muddy bottom. Because the pop sound comes cracking out of the blue rather unexpectedly, the originator of the sound remained a mystery to me for a long time. Only after a few trips in the company of a marine biologist Nicolas Pilcher did I finally learn what the unique loud explosion was. It is produced by the rarely seen mangrove snapping shrimps, which inhabit the muddy, soft substratum of the mangroves. It took me a few more trips to eventually see this stealthy creature in action. You are most likely to hear a snapping shrimp before you see one. They belong to the family Alpheidae. Two species are common in Malaysia: the smaller banded mangrove snapping shrimp, *A. euphrosyne*, which is about two to three centimetres long; and the giant mangrove snapping shrimp, *A. microrhynchus*, which can grow up to eight centimetres. *A. euphrosyne* is the more common species, found in burrows under logs and roots. At low tide on mudflats especially, you can hear the incessant pops of these little translucent-bodied creatures. The large ones can produce a very loud pop. During the day, they are often tucked away in their burrows or other hiding places such as under rocks and mangrove roots. They are more active at night when they forage outside their burrows.

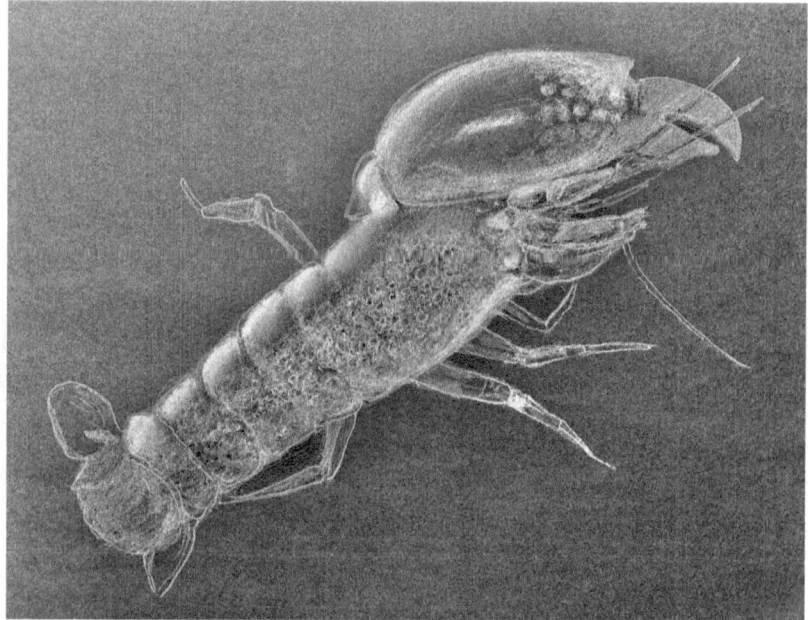

The translucent-bodied mangrove snapping shrimps are found in burrows under logs and roots of mangroves during low tide. Their presence is indicated by the incessant popping sounds coming from the muddy substratum, where they lay hidden in their homes during the day. As the day gets cooler towards the evening, they become active and come out to forage for other crustaceans and bivalves.

I am sure in their previous life, mangrove snapping shrimps were physicists. The mechanism used to work their greatly enlarged pincers seems to derive all the power and energy from the fundamental principles of physics. The dimension of its disproportionately oversized pincer may be as long as its entire body. With this, it can produce a loud explosive sound beyond imagination, an unbelievably loud pop blast from such a diminutive creature. More interestingly, the resultant explosion is so powerful that it can stun prey such as fish or even crack open hard shells of small clams. I was absolutely amazed to finally understand how such little crustaceans can have created a sound so loud and powerful as to paralyse a fish many times larger than itself. Indeed, to understand how that mysterious sound was produced actually necessitated some recollection of my elementary high school physics. Its large claw is packed with muscles, which acts like a moveable 'finger' held at right angles to a matching 'socket' on the opposite claw. When these huge muscles are pulled, immense elastic force is unleashed, causing the pincer 'finger' to hammer into the socket at an incredible speed and force. This produces a loud explosive sound. The sound, however, is not the result of the huge claw hitting the socket like the sound produced when we clap our hands together. Rather, the forceful snap results in a jet of water travelling at such high velocity to cause a sudden drop of pressure in accordance with Bernoulli's principle. Under such conditions, a tiny vapour bubble is formed, which physicists call a cavitation bubble. Upon exposure to the pressure of the surrounding water, the cavitation bubble collapses with a loud explosive sound. During the collapse, for an infinitesimally brief moment, the temperature in the bubble is said to reach the surface temperature of the sun. Amazing is an understatement! Snapping shrimps employ these underwater explosions to fend off predators. It is interesting to note that snapping shrimps have been the subject of active military research since World War II after their explosive sounds were found to interfere with the detection of hostile enemy submarines!

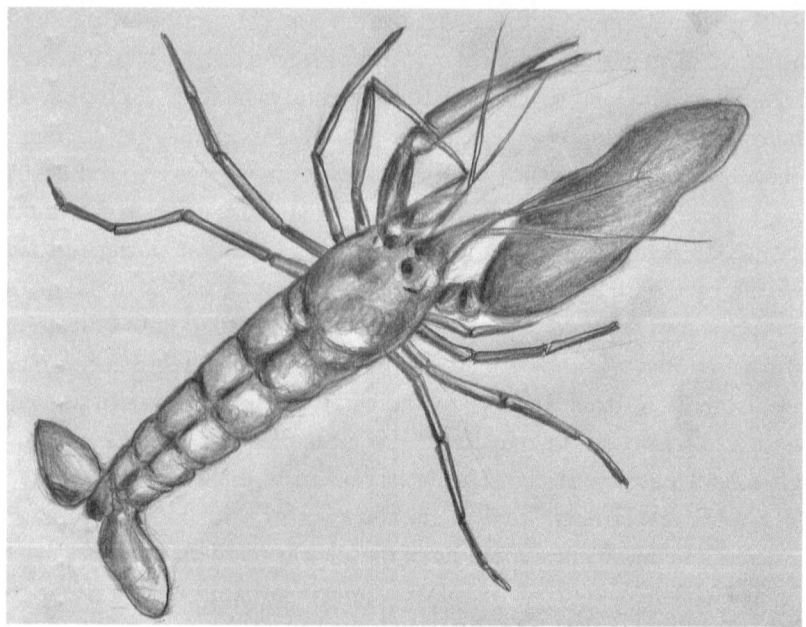

The mangrove snapping shrimp has a huge claw that snaps on a much smaller claw to compress water, producing heat to about 4000°C. A low-pressure cavitation bubble is formed, which expands and bursts, causing the loud popping noise often heard in mangroves.

Scientists believe snapping shrimps also employ their pop sounds to communicate within their own species. What led to this notion is also interesting. Some species of snapping shrimps are known to live in symbiosis with a small goby, a species of marine fish that typically has a sucker on the underside. The goby shares the same burrow with a snapping shrimp in a symbiotic arrangement. With keener eyesight, the goby keeps a lookout for predators, while the shrimp works hard digging and maintaining their shared residence. This relationship is amazingly close and endearing. The shrimp is constantly touching the goby with at least one of its antennae at all times. Literally inseparable, when the goby scurries into the burrow, the shrimp follows behind!

Of all means of communication, visual is almost universal to all animals. It involves signals that can be seen. They can be manifestly conveyed amongst species in gestures, facial expressions, body postures, and colouration. We are familiar with primates raising their arms, slapping the ground, or staring directly at us at the zoo. These are gestures and postures intended as a threat to an intruder. Visible movements are widely used by macaques to communicate for example. They raise their arms, slap the ground, or stare directly at another monkey or human intruder. In the primate world, establishing dominance in social hierarchy or defending territory is also communicated through visual gestures. Amongst themselves, when approached by an alpha male, the young male macaques often give a facial expression in the form of fear grins. This is a signal of submission indicating they accept the male's dominance. In social species, visual communication is also key in coordinating the activities of the group. Dominant males ensure they stand prominently visible to others when marshalling and coordinating group behaviours within their troop. During food acquisition and defence from intruders, they are similarly clearly visible by all members to maintain group cohesion.

Communication using gestures is almost universal in the primates. Facial expressions of threats are used almost in all monkeys especially across the macaque species. It typically involves glaring intensely at the opponent with eyes wide open, gaping mouth with bared teeth, eyebrows raised, and ears flattened.

Another important part of the communication repertoire of mangrove species is tactile signals based on touch. This is only applicable when animals live in proximity to each other. Most social animals living in family groups can number more than twenty individuals. Tactile signals play an additional role in strengthening social relationships. Through constant or regular physical contacts, they nurture and maintain social bonds between related animals and those familiar with the family unit. In this regard, the way humans interact with one another seems much the same to that of monkeys and apes. Our customary practices do not seem alien at all when compared to the many ways they interact. Acceptable forms of tactile communication amongst us have become our normal social culture. Handshakes, hugs, or pats on the back are amongst our means of tactile communication. But animals have far more variations of tactile communications than we do. These can include nuzzling, licking, head rubbing, pawing, body contact, boxing, and biting. Each of these means of communication can express either affection, anger, warnings, or dominance depending on the species. In many primate species, members of a group will spend a good part of the day grooming one another. On closer observation, they are actually removing parasites and performing other hygienic tasks to keep each other clean. In doing so, this largely tactile behaviour reinforces cooperation and social bonds amongst group members.

Tactile communication may have a more profound and long-lasting impact in caring of the young ones. Mothers are often seen licking their newborn profusely in the first few hours after birth. Such tactile communication is believed to stimulate breathing and blood flow of the baby. At the same time, she is cleaning and transferring her scent to a new member of the family. Newborns will respond to their mother's nuzzling and licking by moving closer into mother's cuddle. They also lovingly rest with their bodies touching the mother's chest to share body heat. Such cuddly responses to mother's tactile communication will also better protect themselves from predators, which can be nearby ready to pounce on the helpless vulnerable newborns.

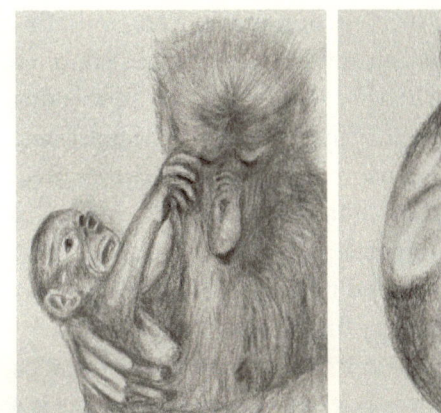

Tactile signals through touching and grooming play an important role in strengthening social relationships amongst higher animals like primates. Through constant or regular physical contact, they nurture and maintain social bonds between related individuals and those familiar with the family unit.

Amongst many mammals, maternal care strongly impacts infant survival. The influence of parental presence during infancy can accelerate offspring sexual maturation and fitness. In parental care, silver langurs are adept at cooperating and solving problems collectively. Such cognitive awareness is shown in alloparenting, whereby an infant is cared for and raised till weaning by not just the mother but also other unrelated adult individuals.

As previously stated, the noisy backdrop of a mangrove forest is often too overwhelming and constantly interfering with the auditory communication of animals in the ecosystem. Under such circumstances, communication amongst species, which rely primarily on sounds to be connected, often become difficult. This can adversely affect their chances of meeting partners to mate. In the long term, their species abundance declines, and overall survival of the species in the mangroves is hugely compromised. There is, however, a way around this. Some species have evolved to communicate using chemicals or pheromones as signals. Detection of these chemical signals by smelling is not affected by the presence of drowning background noises. This is called olfactory means of communication. An excellent means of sensing odours can maximise interactions between species to carry out basic life functions. Selection of superior clones is a matter of survival to many species living in a noisy environment. It helps during foraging because they need to smell and locate the sources of food. Predators emit recognisable odours that can be detected through smell. This can assist the potential victims sense a predator's presence and make a timely escape. More importantly, animals using the olfactory sense increase their breeding success because they are not struggling to detect the smell of partners ready for copulation amidst the deafening background noise of the mangroves.

In many species, the evolutionary and adaptation processes have resulted in their visual, olfactory, and tactile senses being more powerful than others in the same ecosystems. Their ability to see, smell, and touch greatly differ. One or more of these senses have evolved to become more superior and sophisticated in serving their respective species. A combination of all visual, olfactory, and tactile communications is common in the insect world. Social insects seem to have evolved an ingenious means combining the three senses—namely, visual, smell, and touch—to the best advantage in their social lifestyle. To stay connected, ants and bees use their eyes to detect visual signals and a pair of antennae to smell and touch. Despite having a pair of compound eyes, ants and bees possess antennae, which allow them to stay in touch amongst individuals of the same species as well as with their surroundings. The antennae are dotted with sensory cells capable of touching and communicating with each other. Ants and bees navigate, detect odours, taste, and even hear with their antennae. Like all sense organs, these antennae send the information they gather straight to the brain. They secrete a chemical signal, pheromone, to trigger a response in other individuals of the same species. Pheromones are especially common amongst social insects, such as ants and bees. Its function is multidimensional. The chemical serves to attract the opposite sex, raise an alarm, mark a food trail, or trigger other more complex behaviours.

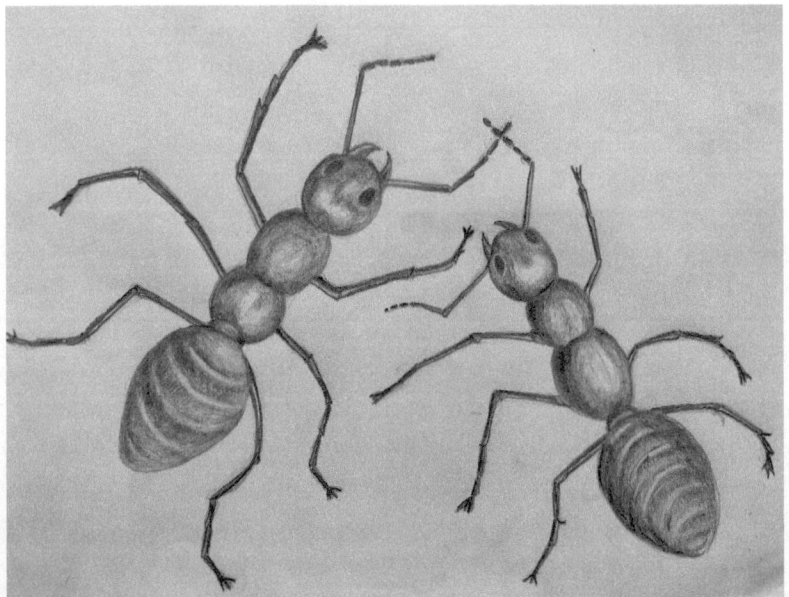

Invisible to our naked eyes and inaudible to our ears, ants communicate using chirping sounds called stridulating by rubbing two parts of their abdomen together. They also communicate using signals with their legs and use their antennae to sense different chemicals, which other ants lay on the ground called pheromones.

Ants seem to use pheromones superbly in their communication despite their ultra-busy lifestyle as social insects. They leave a scent behind on their trail wherever they go so that others in the vicinities looking for food do not get lost returning to the nests with their collection. Everyone will be able to find their way home by sniffing the scent. They also lay down pheromone trails to direct others in the colony to newly found sources of food. When a food source is abundant, ants will call for reinforcements. They deposit pheromone on both the outgoing and return legs of the journey, marking the trail and attracting more ants. When the food source is about to run out, the ants will stop adding pheromone on the return, allowing the trail to fade out.

Ants also have the ability to organise themselves into 'castes' or a social status, each playing its specific role in the colony. Remarkably, each 'caste' responds differently to the same hormone signal. A squashed ant, for instance, would release a burst of pheromones capable of alerting other ants of danger in their proximity. In challenging a perilous intruder, they will seek help from their own species. Exceptionally, more hormones will be secreted to incite more ants swarming and stinging the intruder aggressively.

In addition to olfactory means, social insects also have amazing tactile communication. Upon finding food, a honeybee on a foraging trip will perform an elaborate sequence of motions called a waggle dance. Such energetic boogie-woogie moves indicate the location of the food source. Since this dance is usually performed inside the nest in total darkness, other bees will interpret the dance primarily through tactile response by touching the dancing messenger all over the antennae and legs. Some insect antennae have evolved to become super-specific in function. Mosquitoes have feathery antennae that respond to just one sound—that is, the beating of a female mosquito's wings. The bees have segmented antennae that are super-sensitive to floral scents. They are able to sense and choose the perfumes of thousands of different flowers. Moreover, studies suggest bees also use their antennae to measure humidity and flight speed, a type of speedometer. Flies, however, do not have long antennae. Instead, they have sensory hairs covering every part of their body. Those found on their feet and mouth are used to taste. If the fly walks on something edible like sugary foods, only then will they lower their mouthparts to take a bite.

Antennae of insects are utilised not just for smelling the chemicals in the environment but also to feel the texture and hardness of object surfaces around them, sense the temperature of food, listen to sounds emitted by their own species or predators, and detect the movement of air or wind.

Bizarrely, the functions of antennae extend beyond sensing to smell and touch. The antennae of a longhorn beetle can grow longer than its own body. During copulation, these long antennae are their loving 'arms', which are useful to embrace and hold its partner steady. Longhorn beetles also use their antennae under fierce and aggressive situations. They use them during duels when fighting with other males for mates. They will use their long hardy antennae in the combat, which can end up toppling the competitor on its back or it would walk away exhausted.

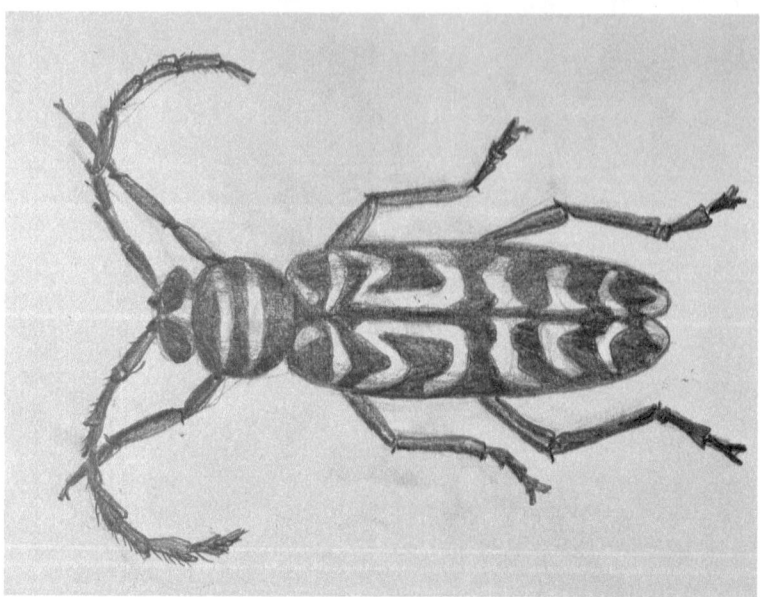

The longhorn beetles use their antennae to smell and touch as well as sense. Their slender pair of antennae, which can be longer than its own body, functions as their loving 'arms' during copulation to embrace and hold its partner steady.

About fifty metres along the bank, the larger trees promised some cool shade and drier ground. I fought my way ahead to take a break and put my feet up for a while as my strength had begun to wane. I heard familiar sounds of geckos barking in the background. The tokay gecko, *Gekko gecko*, is a nocturnal arboreal but can be heard calling for mates just before daylight disappears. This native to Asia and some Pacific Islands is found inhabiting rainforest trees including the mangroves. They can easily adapt to rural human habitation, often on walls and ceilings at night, feeding on small insects attracted to electric lights. Endowed with relatively wide foot pads, it is a strong climber and able to support its entire body weight upside down or on a vertical surface for a long period. This big-eyed, soft-bodied reptile may appear supple and friendly enough to grab and place one on your palm. But be warned: this sad-looking creature is generally territorial and aggressive. It is known to inflict a strong bite when handled.

Amongst the many talents of geckos is their ability to scurry along seemingly impossibly surfaces upside down, such as glass windows or across ceilings. They have specialised sticky toe pads lined with more than six million nanoscale hairs known as setae. Some geckos have the ability to 'fly' through trees, change colour, and even call to find a mate by barking.

In the wild, the male gecko's mating call is a loud repetitious croak sounding like 'tok-kek, tok-kek, tok-kek' from which both the common and scientific names were derived. It also barks when threatened as it takes a defensive posture with its mouth wide open. Interestingly, different cultures have interpreted the barking sounds differently in Asian countries including 'gekk-gekk' and 'pook-kay'. The US Marines fighting the Vietnam War heard tokay sounds as 'tuck-too' and assigned the moniker 'fuck you' lizard to this harmless creature.

Clearly, hearing is important to survival in noisy mangroves. To circumvent these constantly distracting noises, animals must evolve super-efficient hearing ability. Geckos generally have a small circle of discoloured skin by the sides of their head. These are eardrums, which serve to pick up vibrations from the air and transmit them into the middle ear cavity. Some animals appear to have no ears at all. But the truth is their ears are hidden in some form or another. They have evolved their own clever ways to make sense of the sounds around them especially if they depend on hearing where their food is coming from. Spiders hear through their legs. They detect vibrations, which make the hairs of their legs wave and sway. A message is when food is caught and is struggling to break free from the web. Insects detect sound through thin membranes, which can be anywhere on their body. The grasshoppers use their legs and abdomen. Male grasshoppers and katydids sing by rubbing two parts of their body together. Most species rapidly rub a rough 'file' on the inside of their back legs against hard ridges on the edge of their wings. The vibrations of specific messages are picked up using hearing organs found on their abdomen. Each species has its own unique series of chirps or clicking sounds recognised only by the females of their own species. There are three types of songs: one to invite a female to come closer, second to lure and persuade her to mate, and another a battle cry warning other males to stay away.

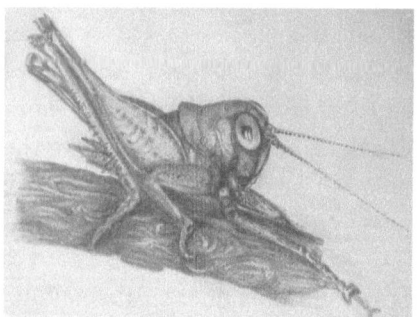 

Grasshoppers and katydids produce sound by stridulating, an act of rubbing two body parts together. It can be likened to playing a violin, where the bow is drawn over the strings to create friction, thereby generating acoustic vibrations. These little creatures have a series of small pegs on the inside of their hind legs that are rubbed against the wings to create sound waves, conveying specific types of information to be detected by members of the same species.

Hearing is not only useful in search for food but also important to the mating success. Calling males must be able to hear the replies of willing females in the noisy mangroves. The calls ought to be pitched at the right frequencies, which are low or high enough for their mates to hear. For vulnerable insects, this is important so as not to increase their chances of being heard by predators. Some animals have evolved to emit low-frequency calls, which travel longer distances to solicit mating partners. Mice communicate with one another at high frequencies to alert one another of danger or woo a would-be mate. Sound waves travel more easily in water than air. Fish that live in water do not evolve any external ears but have inner ears. It has been documented that even walking fish of the mangroves, the mudskippers, scream at each other when they are out of the water. Mating partners will produce both pulsed and tonal sounds of low frequency during their encounters.

My many enjoyable explorations of the tropical mangroves have given me much deeper insights of the many trump cards biological evolution has played on the table. Indeed, myriads of successful clones have been created in this game. Without a doubt, staying connected is crucial to animals. To us humans, staying connected to one another can be optional. But to mangrove species, it's obligatory, absolutely mandatory for their survival. They do this through communication using an array of signals including visual, auditory, olfactory, and tactile. Their means of communication have been adapted behaviourally to assist them staying connected in finding mating partners, establishing dominance, defending territory, coordinating group behaviour, and caring for their young.

10

Good With Food

To have survived to the present day, the path to obtaining food by our ancestors couldn't have been smooth. Their hunting and gathering methods must have been periodically crippled by insufficient and irregular food availability. Changing seasons and rapid degradation of food were something of insoluble consequences. Having just to survive was difficult. At times, they had to bear the brunt of famines and diseases, causing calamitous scarcity of food. To successfully navigate an uncertain future, humans began by learning from their past experiences in obtaining sufficient food. The Neolithic Revolution, also known as the Agricultural Revolution, is thought to have begun about twelve thousand years ago when man started to cultivate food through farming. Since then, we have progressed to produce the vast majority of the world's food supply.

On my many explorations of the mangroves, I seem to have encountered different wildlife species on each visit. It was dependent on the time of the year. The sightings of wildlife were more frequent during the rainy season when the trees were lush green with new leaves. These same animal species were rarely encountered during the dry season when fruit and young shoots were scarce. Wildlife distribution clearly depends on the availability of food in the mangroves. For most of the year, the abundance of food dictates the presence of mangrove wildlife, which come out to forage and reproduce. The population of exclusive residents to this ecosystem like the proboscis monkeys might not see a significant decline because they are highly adapted to living in the mangroves during good and bad times. Dearth of their special diets doesn't significantly affect their activity and survival. They have become accustomed to finding alternative sources of food that can sustain them all year round. But the number of mangrove visitors like silvery langurs, long-tailed macaques, or small-clawed otters may change in response to the availability of food. These opportunistic species appear to only visit the mangroves when food is plentiful and readily available.

The sky above was blue with patches of showy white clouds. Not too far from the coastlines, dark silhouettes of water huts for mussel farming dotted the horizon. From the air, a voice of nature stole my attention. It was a bird of prey, white on the head, rump, and underparts but dark grey on the back and wings. In flight, the black flight feathers on the wings are easily seen from below. I heard a familiar loud call reminiscent to that of a honking goose. *It must be the breeding season for the white-bellied sea eagle, Haliaeetus leucogaster*, I thought, having heard similar calls on my last trip to Pulau Perhentian, off the east coast of Peninsular Malaysia previously. Evolution has been kind to this bird with respect to finding food. Designed to be skilled hunters from the sky, they are equipped with a large hooked bill and feet with long black talons. They are a common sight in coastal and interior wetland areas of Malaysia. Their loud 'goose-like' honking call is a familiar sound, particularly during the breeding season. Often sighted perching high in trees, they soar gracefully over waterways on the search for food. They feed mainly off aquatic animals, such as fish, baby turtles, and sea snakes. Skilled hunters, they have been reported to prey on birds and mammals as well.

Evolution has endowed the white-bellied sea eagle every conceivable attribute to make it a highly skilled hunter from above. They have huge strong wings to gracefully uplift themselves and fly through the air seamlessly, as well as excellent eyesight, a hooked bill, and feet with long talons for seizing prey while in flight.

Another graceful bird in flight came into view. Soaring high above against the cloudless blue sky, I strained my eyes in an attempt to identify the magnificently handsome winged creature. It was the Brahminy kite, *Haliastur indus*. Another graceful bird in flight, it is a common bird found throughout much of India and Southeast Asia as well as Australia, wherever there is water of any kind. The adult bird is easy to distinguish, having reddish-brown plumage and a contrasting white head and breast. The name was bestowed upon the birds by Europeans when they learned that this kite was sacred to the Indian god Vishnu. In reality, the lifestyle of this bird is far from sacrosanct or holy. It is a scavenger extraordinaire. It will eat virtually anything meaty and is particularly fond of fish. Watching Brahminy kites hovering and hunting their prey has always been a delightful experience for me. Upon spotting their prey, the kites would dive feet first instead of head first and seize the victims with their claws. Jubilant, they would elegantly swoop skyward and squeal out distinct high-pitched calls that could be heard for miles as if announcing to the whole world of the grand catch.

Besides binocular vision, clarity of eyesight is another incredible visual sense of birds of prey. Spotting moving prey on ground from mid-air can only be achieved with impeccable vision, especially those well-camouflaged against vast open backdrops. Looking for potential mates is less problematic through hearing and recognising characteristic sounds emitted by females. We can see stars glowing from millions of light-years away if our sight is not obstructed. But our vision pales in comparison to that of some animals. Eyesight is more important to them than hearing, smell, or touch. All birds of prey have excellent long-distance vision. Compared to humans, they can see clearly about eight times as far as we can. If we swap eyes for an eagle's, humans would be able to see an ant crawling on the ground from the roof of a ten-storey building. We could make out the pains a rugby player felt underneath a scrum through his facial expression from the worst seat in the stadium. Everything that appears directly in our line of sight would be significantly magnified and brilliantly coloured in an unimaginable array of shades.

Eagles and other birds of prey can also see four to five times farther than the average human can—meaning, they have 20/5 or 20/4 vision under ideal viewing conditions. They owe this incredibly sharp vision to two anatomical features of their pair of eyes. Both facets are different to ours. Firstly, their retinas are more densely coated with light-detecting cells called cones, enhancing their power to resolve fine details. This is equivalent to a higher pixel density, which increases the resolving power of cameras. Secondly, they have a much deeper fovea, a cone-rich structure located behind the eyes. All animal eyes have this cone-rich fovea, which detects light coming through the centre of the visual field. But our human fovea is a little shell or bowl, while in birds of prey it is a convex pit. Investigators think this deep fovea allows their eyes to act like a telephoto lens, giving them extra magnification in the centre of their field of view.

Thanks to the superior anatomy of their eyes, birds of prey can see four to five times farther than humans. Their incredibly sharp vision is due to their eye retinas, which are densely coated with light-detecting cells called cones, enhancing their power to resolve fine details.

In addition to the sharp focus and a central magnifier, birds of prey see in brilliant colours of all shades, permitting them to detect unbelievably small changes in the colouration of their prey. This is true of all birds. They have superior colour vision, enabling them to discriminate vividly what they see much better than humans do. Another stark difference between our vision and that of birds of prey is their ability to also see ultraviolet light. Apparently, they put this uncanny ability to good use during hunting. Scientists believe their ability to see UV light helps them detect the urine of prey on the ground. Potential victims exasperatedly fleeing to escape capture often leave urine trails on the ground, which reflect UV light detectable by birds of prey from the air.

In addition to the ability to see farther and perceive more colours, birds of prey also have nearly double our field of view. Eagles' eyes are angled thirty degrees away from the midline of their faces, permitting them to see almost all the way behind their heads with a 340-degree visual field. Directly, this confers a clear advantage in hunting and self-defence. Humans only achieve a 180-degree field of view. During flight, the eagle swivels its head constantly. To locate prey, this periodic turning of the head to the side enables it to constantly position the fovea at the back of its eyes for a wider field of view. Here again, the fovea serves as a swivelling telephoto lens. After spotting what it is looking for in this manner, instinctively the bird redirects its head towards its target using stereoscopic vision. The brain combines the precise points of view of both eyes to gauge its distance from the target prey. The required speed is calibrated by the brain as it swoops and seizes the prey.

In my mind, I kept recalling the many attributes evolution has truly endowed the birds of prey in enhancing their food-hunting prowess. Like most high-flying predators, white-bellied sea eagles and Brahminy kites have sharp vision, enabling them to swoop down on their prey with impeccable accuracy. Hawks, eagles, and kites possess binocular vision to pinpoint precisely the location of their prey. Unlike most birds whose pairs of eyes are located on each side of the head, birds of prey have their eyes on the front of the head like most mammals. This way, both eyes are able to focus on the same object directly in front. It enables them to perceive depth between themselves and their prey with absolute precision. It is encapsulated in the anatomy of their eyes. Such a pair of eyes with all the sophistication nature could possibly design is crucial if one has to locate food from as far as forty metres from mid-air. I convinced myself of nature's remarkable endowment to all birds of prey.

But there is a downside to this extraordinary gift of evolution: birds of prey such as eagles, hawks, and kites tend to have a greater proportion of their brain size devoted to visual processing than other groups of animals. Scientists believe this comes at the expense of them having less developed senses of smell or taste. Lacking the ability to smell or taste is in fact universal in the avian kingdom. But birds are known to have good cognitive processes. They seem to have areas of the brain that function like the human brain cortex being the part responsible for memory, language, and complex thoughts. Many birds have superb memory. They have the ability to solve problems not easily matched by many mammals.

A reddish-orange dragonfly just flew past. Another player in the game of clones that never ceases to amaze me. This winged creature first evolved some three hundred million years ago. Their ancestors were revealed in fossil dragonflies possessing wingspans exceeding two feet. Today's dragonflies have wingspans of two to five inches only. They are, however, expert fliers. Not only are they able to fly straight up and down but they can also hover like a helicopter. Like birds of prey, they are excellently equipped to seize prey while flying. They can even mate mid-air. Their hunting mode is so efficient that they are almost 90 per cent successful in catching their prey in flight. This superb adaptation has inspired aeronautical engineers to study the flight of dragonflies in an attempt to innovate robots capable of achieving similar feats in mid-air. Additionally, almost the entire head of the dragonfly is but a pair of humongous compound eyes. This gives them incredible vision encompassing almost every angle except directly behind—a useful capability to have in hunting for prey.

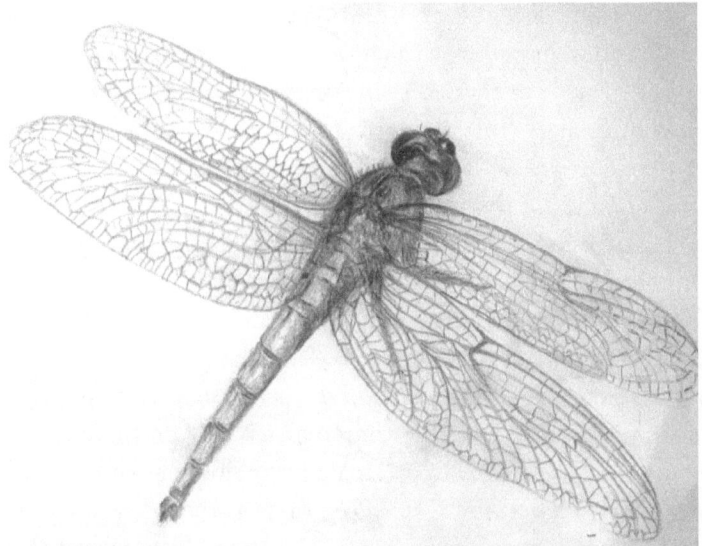

The vision of dragonflies appears tailored to make them excellent hunters of predators. They can detect prey travelling at high speed, enabling them to intercept the prey in mid-air. Their binocular vision enables them to judge distances almost to pinpoint precision.

Moths and butterflies have evolved such that almost every part of their body is capable of detecting odorous chemicals. They also have taste buds located at the end of the tongue and can identify the chemicals using sensory structures on their feet.

In a scrabble game once, I took exception to the word *toxophilite* placed on the board by another player. 'Such a word doesn't exist,' I challenged him, but I lost. It turned out there is such a word, which refers to both the bow and arrow in archery. Interestingly, it was derived from the Latin word *toxicum*, meaning 'poison for arrows', and the Greek word *philos* for 'loving'. This came to mind as I was walking along the wooden walkway constructed in the mangrove forest in Setiu, Terengganu, Peninsular Malaysia. I was a visiting professor at the Universiti Malaysia Terengganu (UMT) at that time. Research scientists Evan and Norhayati were showing me some of the wonderful research projects the staff at the university were engaged with in this mangrove reserve. I heard the occasional soft explosive sounds coming from the surface of the water on each side on the walkway. The soft popping bursts seemed to originate just below the branches of leaves hanging amongst a maze of *Rhizophora* roots as the tide was coming in. I was told they were sounds of archer fish hunting. The archer fish is from the family Toxotidae, known for their habit of preying on land-based insects by shooting them down with water droplets from their specialised mouths; hence the common name archer fish and scientific family name Toxotidae from the Latin *toxicum*.

This remarkable fish has an elongated body that is laterally compressed, making it appear slim at approximately twenty centimetres in length. It has a pointed face markedly narrowing into a snout around the mouth. There are seven different species of archer fish, which vary with the darkish band patterns found on the body of each species. The most common is the banded archer fish. Their body is silver with a gold tint across the back and four to six vertical bands across the body.

Looking like any other fish, the accentuated anatomy of its pointed mouth enables archer fish to take scrupulously accurate aim when shooting down their prey above water. It acts like a powerful toy weapon, the Super Soaker water blaster, children use in swimming pools.

While the archer fish may look prosaically ordinary like other fish, the means by which it captures prey will stop you in your tracks with amazement. Stunning, almost literally! The accentuated anatomy of its pointed mouth allows the archer fish to take scrupulously accurate aim when shooting down prey. It is equipped with a powerful weapon, which works like the Super Soaker water blaster my grandchildren run around with, wetting everyone during barbecue time in the backyard on hot summer days. Armed with a similar operational mechanism, the archer fish is able to fill its mouth with water and fire streams of water to hit the prey.

Archer fish can be found in estuaries with brackish water around mangroves. They are excellent hunters. They spend the majority of the day swimming just below the water surface looking for food above water. With its slim body, the fish silently float near the surface looking for spiders and insects. Once a tasty-looking bug is directly above, it will shoot a powerful stream of water to knock the target into the water. They have been observed shooting at prey 1.5 metres above the surface of the water. When preparing to shoot the prey, they suck in water, place their tongue on the roof of their mouth, and then shoot. If not successful at the first attempt, they repeatedly aim and shoot up to seven streams with one mouthful of water. Should that still result in a miss, the fish will use its final hunting skill by jumping out of the water with enough momentum to grab its meal directly.

The archer fish hunts insect prey by shooting a stream of water, which knocks the prey from branches into the water where they can make be eaten. As if that's not smart enough, archer fish can also recognise human faces, a feat that is thought can only be accomplished by primates and birds, which have a large and complex brain.

Archer fish must have excellent eyesight to succeed in this mode of hunting. Indeed, they do have more developed eyes than most fish. The binocular vision allows them to focus forward and judge longer distances. While its mouth goes above the water surface to shoot down prey, the rest of the body remains submerged. There is a remarkable action in performing this feat. There is a fundamental principle in the physics of light to overcome. This is a phenomenal adaptation in itself. The eyes of vertebrates usually see well in water or air. Human eyes evolved to see best in air, while fish see best in water. Human vision is adversely affected in water, and similarly fish cannot see very well outside water. As dictated by the laws of physics, light rays bend when travelling from one medium to another, and the amount of bending is determined by the refractive indices of the two media. When light enters human eyes, these bent light rays entering our eyes can cause blurring of vision if not corrected. Human eyes are able to focus these light rays on the retina at the back of our eyes, resulting in the clear focused images we see. Water has a significantly different refractive index to air. Fish adapted to see clearly underwater are not able to focus on images of anything above water. How does archer fish make the adjustment in its eyesight to compensate for the refraction of light that enters the water from above? The archer fish must learn to aim and shoot its above-water target with absolute accuracy or go hungry. This superb ability remains vague until today.

A recent discovery in a small tribe of sea nomads living along the west coast of Thailand is interesting. For centuries, these nomadic people have lived on the islands of the Andaman Sea, harvesting clams, sea cucumbers, and other marine morsels for food. They are excellent divers. They are able to gather food by plucking their fare off the seabed as deep as seventy-five feet. But what is most impressive is their underwater vision. Without goggles or other diving aids, sea gypsy children routinely spot even the smallest of shellfish at the bottom of the sea. Studies have shown that their ability to see underwater is an adaptation not determined by genetic or anatomical features. With little practice, their unique vision can be acquired by any young person. The children of sea gypsies have better than twice the underwater resolving power of European children. Such level of underwater acuity was previously thought impossible in humans. A study suggested how this was achieved. The children did not flatten the corneas on the front of their eyes like some amphibious birds, fish, and frogs do to improve underwater vision. Rather, they shrank the size of their pupils, the round black aperture through which light enters the eye. Underwater, the diameter size of their pupils is reduced to 1.96 millimetres, which is 22 per cent smaller than the 2.5 millimetres minimum seen in Europeans. This extreme reaction, routine in the children of sea gypsies, is completely absent in European children. On the contrary, the pupils of most people in fact enlarge slightly underwater in response to the lower level of light. The study showed the children can train themselves to constrict their pupils when diving and enhance their underwater visual acuity. The fact to note from this study is enhancement of visual acuity is a learned behaviour. It is an adaptation not simply inherited as an inborn reflex. Extending this observation to the case of archer fish, it would not be unreasonable to suggest the ability to see its prey outside the water is also a learned skill. Their young must start learning how to shoot water out of their mouth very quickly in life, or they will go hungry. This seems to be true because even as young fries, archer fish are able to shoot jet streams of water up to about eighteen centimetres.

Sometimes, the idea of exploring the mangroves on a night walk seemed attractive. But the stillness and cooler night air of the mangroves failed to offer any respite. The tropical setting of almost 100 percent humidity made perspiration a resultant feature. After only twenty minutes of moving, I was drenched in sweat. I slowed down, allowing for a good inspection of things around me by shining my headlight through and behind the thick foliage further ahead. Suddenly, I caught sight of a lump of yellow stripes still and glistening on a low branch against the pitch-black backdrop of the night. It was a distinctively coloured mangrove yellow-ringed cat snake, *Boiga dendrophila melanota*, having an evening nap. It was coiled amongst the branches hanging over water gradually ebbing to sea. On closer examination, the coiled mound was in fact mainly black, with narrow yellow bands along the body length. Mangrove snakes are considered nocturnal, rarely seen hunting during daytime. They sleep or rest during the day. When night comes, they set out to ambush their prey, primarily targeting reptiles and amphibians including lizards, frogs, and toads. They also search and stalk for sleeping birds and eggs. Sometimes, rats and birds fall victim to larger adult *Boigas* also. It can slither along very fast. On one of my night outings, I have seen this species speeding in hot pursuit of a fumbling rodent within the complex network of mangrove roots. *Boigas* are only mildly venomous. But its colour pattern was reminiscent of the deadly venomous banded krait, *Bungarus fasciatus*, also a known resident of the mangroves. A case of mistaken identity could spell fatality to the uninitiated, so I decided not to venture closer and leave a sleeping snake to lie!

Yellow-ringed cat snake has a super-sensitive sense to help locate and hunt prey by heat. Unlike the mangrove vipers and kraits, which use similar heat sensor pits located behind their throats, cat snakes swallow their victims whole and break them up internally using bony protrusions in their throats.

Slithering stealthily and hissing as they make their way on the branches, snakes understandably make many visitors to the mangroves quake with fear. But I have always found the yellow-ringed cat snake attractive and pleasant to watch. Some snakes have evolved a super-sensitive sense finely tuned to help locate and hunt prey by heat. They have special heat sensor pits located behind the nostrils. They wait in trees and lash out swiftly with a powerful bite. Unlike the mangrove viper and krait, the cat snakes swallow their victims whole and break them up internally using bony protrusions in their throats.

Technically speaking, mangrove snakes, including all other *Boiga* species, possess dentition described as opisthoglyphous. This means they are rear-fanged, a term referred to a range of unrelated snakes that possess a venom-producing gland at the back of their upper jaws. Their fangs are actually oversized teeth that are elongated and grooved. This is in contrast to front-fanged snakes such as the notoriously poisonous vipers and cobras. These deadly snakes have hollow fangs, which function like hypodermic needles capable of injecting venom when they bite. The delivery mechanism of venom by the *Boiga* is not as efficient as the pit vipers because the fangs are located further at the back of the mouth. Rear-fanged snakes need to securely hold their victims before injecting their less lethal venom. To finally paralyse their victims, rear-fanged snakes need to chew and massage before injecting enough dose of the venom into the areas they bite.

The fangs of *Boigas* and constrictors such as pythons are located at the back of the mouth. They need to securely hold their victims and chew and massage before injecting enough dose of their mildly toxic venom into the areas they bite.

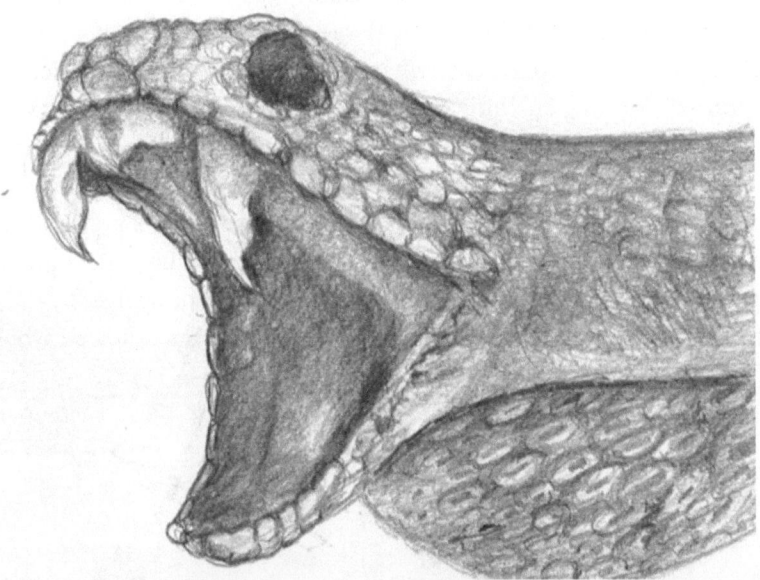

The front-fanged snakes like the notoriously poisonous vipers and cobras are deadly. They have hollow fangs, which function like hypodermic needles capable of injecting enough venom when they bite to kill their victims quickly.

'Why are they called cat snakes, then? They would definitely not go for cats as food. Wild cats would be too big and fast to become prey to this slithery snake,' a student once enquired on one of our field trips. A head-scratching moment followed. Silence and confusion reigned. But trust Professor Indraneil Das, who was functionally a walking encyclopaedia on all snakes and frogs. They are his love and passion. 'Actually, the name cat snake didn't come from the food this snake eats. It came from its eyes. The snake's eyes are large with narrow vertical slits like cat's eyes, hence the name cat snake,' explained Neil. 'This is an eyesight adaptation for nocturnal species. Such eye structure allows it to see better at night when it is actively on the hunt.' I trusted his explanation, but it didn't stop me from thinking. I decided to be a smart alec as well and sprung up a surprising quiz, 'But frogs' eyes have narrow vertical slits too, don't they? Why isn't there a similarly named group called cat frog, then?' Silence reigned for the second time. But no one took my question seriously.

The largest snake often found in the mangroves is the reticulated python. It is known to grow to lengths of more than seven metres, arguably the longest snake in the world. Contrary to what most people assume, constrictors like pythons kill prey by crushing them. The python's most important weapon is the powerful muscles all along its length, enabling it to crush its victim and consume its meal. As a matter of fact, pythons kill victims by giving them a death hug. They wrap their long body around the prey and utilise their powerful muscles to slowly squeeze it to death. The snake tightens its hug every time the victim breathes out. It patiently continues strengthening the grip till the victim suffocates before the victim is swallowed whole.

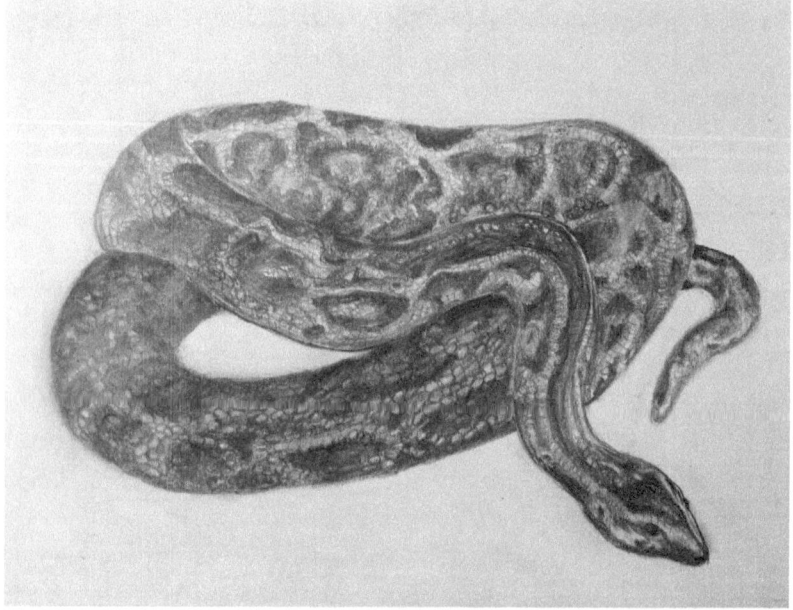

Python ambushes, wraps itself around its prey, and suffocates the victim within minutes. It doesn't devour its victim externally but swallows it whole even if the food is many times larger than its head. Their jaws are connected by very flexible ligaments, so they can stretch around large prey.

The amazing fact about the python and its pursuit of food is that it never gets deterred by a potential victim many times larger than its mouth. Size is no reason for rejection. Nothing that moves and is meaty is too big for a hungry python. It can stretch open its mouth up to 150 degrees to swallow its prey. This extraordinary gape is achieved from having an elastic ligament connecting its jaws. Because the elastic ligament stretches and extends quite substantially, a snake in the midst of swallowing its oversized victim appears as if it has dislocated its jaws. It will always swallow its victim head first. This is made possible by having its teeth facing backwards, which helps to hold securely onto the struggling victim.

If the victim happens to be a large animal, digestion can take weeks or even months. They are often encountered in a non-active state coiled up sleeping on mangrove branches or amongst the shady undergrowth. Sunlight is believed to help the digestion process for snakes and other cold-blooded reptiles. The sun's rays provide the animal with the heat and energy needed for digestion of the victim it swallowed and for absorbing the nutrients from its meal. So, when you see a python coiled comfortably enjoying basking in the sun, you can be rest assured it will not be interested in having you as its next meal. That is a well-fed python.

Pythons are not alone in their ability to pursue victims much bigger than their mouths. I was drawn to a scenario in the mangroves where an army of ants was seen hauling their oversized spoils. It was a dead grasshopper being hauled away by hundreds of tiny ants along a mangrove branch. How could a grasshopper, despite its humongous size next to an ant, be defeated by tiny ants? These were army ants known for their absolute mastery in wrestling insects hundreds of times larger in size as food. They are vicious insects that live in colonies of millions capable of sweeping across vast areas, destroying every living creature in their path as long as they are acceptable for food. Their victims are defeated by their unstoppable aggression and overwhelming numbers. Hopelessly outnumbered, their victims will be killed by fierce bites and venomous stings before being dismembered to help feed the millions of individuals in the colony. The same weapons and battle strategy are employed against intruders and large predators threatening to make meals of them. Despite their size, predators are no match for an aggressive colony of army ants. The victim is doomed the instant it scuttles across the path of a marauding militia of ants. I have seen lizards and skinks being hauled away in the same manner. Bizarrely, army ants are essentially blind. How do they know when to attack? It turns out the attack is triggered by chemical signals emitted in increasing concentrations during the battle. The increase in the concentration of chemical signals results in intense ferocity and aggressiveness of the attack.

More interesting is the way these ants move their prized catch back to their nest. They do this in a well-coordinated manner involving division of labour in which all participate. As they move, they improvise bridges with their bodies by latching on to each other with their mandibles or mouth parts clawing their way back to the nest. It's all-for-one-and-one-for-all attitude all the way.

Almost right before my eyes, I could see the tide steadily rising. Against the blazing evening horizon, convoys of bats were seen gliding and splicing in mid-air feeding. I felt bad for these creatures. In all cultures throughout the globe, bats are long overdue for a major image rebranding. Its oversized ears, curvy obnoxious-looking snout, and toothy smiles make this flying mammal an unlikely YouTube star. There is even a bona fide syndrome called chiroptophobia for a person with a persistent fear for bats. Public relations have never been good for the bats. I feel sad we often fail to appreciate the benefits we receive from bats. The bats save us billions of economic worth every year by eating insects that could have inflicted devastating diseases and destroyed our crops. Disease vectors like mosquitoes are on the menu of bats. Every evening, literally tons of mosquitoes are consumed by millions of bats that fly out of Mulu Caves in Sarawak. Durians and bananas are amongst the tropical fruit that almost entirely depend on bats for pollination. Bats are perhaps more of a man's proverbial best friend than dogs.

A swooping bat above your head is more likely to be hunting mosquitoes than heading for you. How they catch insects while flying in the darkness of night is one of the epic wonders of nature. Bats are the only mammal that can fly and use ultrasound to navigate through echolocation. When sound waves travel and strike an object, they are bounced back. This reflected sound is called an echo. Humans do not hear echoes often, but bats have evolved to hear ultrasound and precisely interpret these echoes. This echolocation has become their primary means for finding food or avoiding predators.

Bats are the only true flying mammals, quite apart from the flying squirrels, which in actuality glide through the air. To fly, bats possess unique wings consisting of four elongated digits—namely, digits II–V. The wing structure is very much like a human arm and hand, except it has a thin membrane of skin called the patagium extending between the 'hand' and the body, and between each finger bone.

The way they exploit this ability to echolocate is one of the magical wonders of animal behaviour. The science behind this is again physics. Bats generate and receive ultrasound waves while in flight hunting in an open space. They constantly produce ultrasonic sounds, which end up hitting solid objects, be it insects or branches of trees. Their amazing ability to vary the frequencies of these sounds and use harmonics to help pinpoint their prey is paramount in foraging. A bat squeaks about five times per second, with each call lasting ten to fifteen milliseconds. When it detects a potential victim, it instantly increases its call rate as well as decreases the duration of each squeak. Each squeak now lasts for only one millisecond, and the squeaks are emitted at a rate of up to two hundred per second. In physics, the time for echoes to return is directly proportional to the distance from where ultrasound is reflected. The resultant echoes are reflected in real time back to the flying bats. This is what the bats 'hear'. It interprets this constant stream of sound waves, which we humans do not hear. The bat brain is able to instinctively calculate how long the ultrasound takes to return to its original source. At lightning speed, this will tell the bat how far it is from the food or the predator.

Amazing in bats foraging is how swiftly the sound signal travels in air at about 343 metres per second to be processed. An echo-locating bat is able to perceive the sound waves, transform them into nervous impulses to the hearing centre of the brain, and interpret the object the sound waves have struck as food to seize or hard object to avoid.

How swiftly the signal is processed excels almost every other animal on planet Earth! Sound travels through the air at about 343 metres per second. The bat brain is able to calculate how far away the object is by instantaneously calculating the time taken for the sound waves to bounce back. To interpret which object the sound waves have struck, the bat has the ability to perceive sound in the same way humans do. The air vibrations are transformed into nervous impulses to the hearing centre of the brain. The electrical impulses of the brain then inform the bat if it is food or to avoid the predator. To have all this happening and functioning so precisely and swiftly is something no human brain can emulate.

Frugivorous bats, which feed primarily on fruit, are important dispersers of seed and crucial in forest regeneration. Because they feed primarily on liquid, bats tend to take the fruit to a safe perch, suck out the juice, and discard the seeds. Small ingestible seeds will be passed out in the bat's droppings away from the parent trees as it flies to distant places through the forest.

Bats are expert fliers and able to turn in tight spaces with great agility and speed. Here, again, while on flight, bats find their way by using sound. They navigate themselves in darkness to find a variety of food including fruit, fish, pollen, and even blood. Many are frugivorous bats, which feed solely on fruit.

Because they feed primarily on liquid, bats tend to take the fruit to a safe perch, suck out the juice, and discard the seed. Only very small seeds will be ingested and passed out in the bat's droppings as it flies through the forest. They are therefore important dispersers of seeds as well. A reduction in the number of fruit bats can result in the disappearance of many species of fruit trees.

When I ventured into the rainforests or mangroves, I usually made an extra effort at attiring myself appropriately. My clothes are usually light-coloured and made from thin cotton. They ought to be loose and cooling for the humid tropical weather. Most of the time, I would wear long-sleeved shirts and baggy trousers. I was only in my shorts and my T-shirt one day. Silly me, I had inadvertently violated my own dress code before setting off on my mangrove adventure that morning. I had forgotten to put on my leech socks (loose thin cloth socks secured by elastic or cloth cords high towards the knees). They are a pair of baggy socks, which are comfortable and effective in preventing leeches getting in between your toes and feet and ending up crawling elsewhere like your belly, armpits, or anywhere warm on your body. I realise I had become breakfast for leeches when I saw deep-red blood trickling just above my ankle. *Darn, that blood-sucking nuisance again*, I sighed shaking my head. I should have known better. Leeches might be small creatures, but they are a perfect bleeding machine. They have ingeniously evolved for finding and feasting on their only food—blood. Assisted by powerful suckers, they cling securely to the skin as they feast on the blood of their victims. For delicious uninterrupted meals, the leeches produce an enzyme in their saliva called hirudin. It is an anticoagulant that prevents the host victim from forming a clot, which would cut the bloodsucking feast short. The secreted hirudin allows the blood to flow smoothly and in sufficient quantity, filling its entire slender body till it becomes well satisfied. In addition, the saliva also carries an anaesthetic chemical, which renders the area surrounding the bite numb. The victim doesn't feel any itch or pain while the leech is pumping in blood, leaving it to enjoy the meal in absolute peace. No wonder I didn't feel anything amiss all morning around my ankle. The little menace must have been holding on to my skin with its hind sucker, siphoning out blood till it fell off under its own weight.

Leeches feed primarily on blood of mammals found living in the mangroves. For a tiny slimy minion living in a vast forest area, finding its next meal could be a huge challenge. The mangroves are not known to harbour many types of mammals as do the inland rainforests. The density of huge mammals in the mangroves is not as high either. A mind-boggling puzzlement occupied my thoughts as to how leeches could possibly obtain enough blood to feed itself in its entire lifetime. Thanks to the evolution of super-efficient suckers capable of sensing the presence of approaching victims, their suckers serve as sensory organs that can detect body heat and movement of their victims. I have seen leeches extend and 'wave' their slender end on leaves quite high above the ground, waiting for animals to pass by. This is a remarkable adaptation in behaviour. In anticipation for its next meal, the leeches must have learned to hang around at heights from where they can hop onto the skin of the forest animals by using leaves as launching pads.

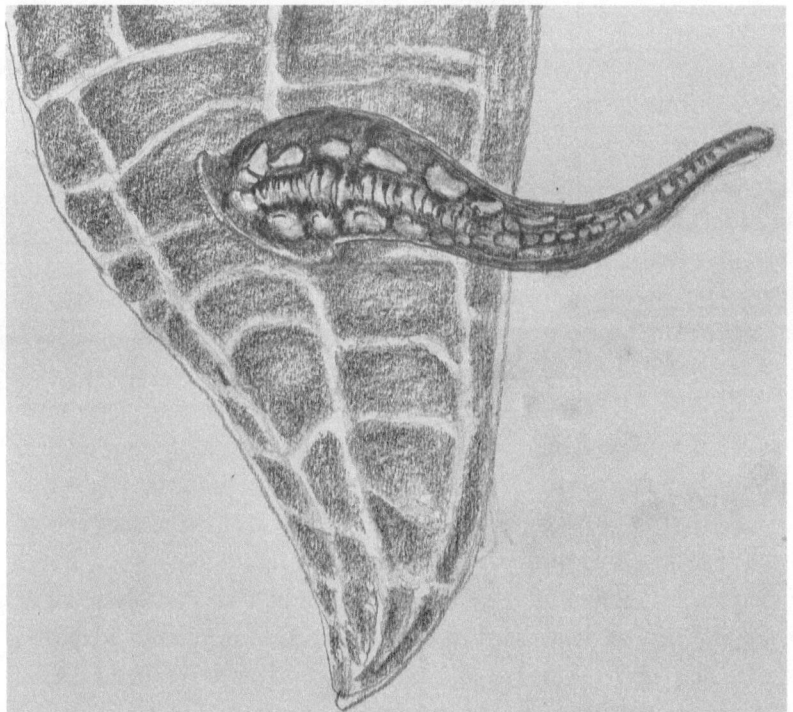

The remote likelihood of latching on to a victim for a blood meal has resulted in leeches exhibiting a remarkable adaptation in behaviour. In anticipation for its next meal, the leeches learned to locate themselves at heights from where they can hop onto the skin of the forest animals by using leaves as launching pads. Here, they would be 'waving' their super-sensitive suckers to detect body heat and vibrational movement of their potential victims.

I wondered if other parts of my body had generously served as blood donors providing free meals to the voracious suckers. I needed to make sure. A nice dry, shady spot along the beach would be ideal and cooling to carry out this inspection, I thought. As I was walking there, I noticed fiddler crabs hurriedly scuttling away on the mudflat. They seemed to have abruptly abandoned their fascinating dance performance and headed for their holes. Somehow, they appeared able to sense and respond to each step I took to get nearer to them. But before reaching their burrows, they seemed to hesitantly retreat in a jerky fashion. They stopped and waited till I took the next step before moving farther away. This was repeated several times till I got too close for comfort when they eventually disappeared into their burrows completely. Clearly, crabs had a system to detect my movements. They have sensitive hairs on the claws and body to detect vibrations caused by my placing my feet on the ground. The same thing happens when vibrations from predators are detected so that crabs are able to make their getaway quickly to safety.

On the contrary, I noticed a starfish nearby that seemed unresponsive to the vibrations I was creating with my footsteps walking towards it. Its five arms radiating out from a central disk-shaped body were hardly moving, but I knew the creature was busy eating. Their arms are equipped with pincer-like organs and suckers, which allow the creatures to slowly creep along the ocean floor looking for food. On the tips of their arms, starfish possess eyespots, which allow them to sense light and dark. This is most crucial in their directional move to their food source. Primarily predatory creatures, they crawl along the bottom of the sea searching and focusing on sluggish or stationary prey within their reach. Favourites on their menu include animal species such as mussels, clams, oysters, and sea snails. They locate their prey using the light-sensing eyespots at their arm tips. Most impressive in starfish adaptation is their feeding mechanism. Because of their tiny mouths on the underside of their body, starfish have adapted an ingenious way of eating items larger than their mouths. How? Starfish can eat outside their body. They have evolved to digest food outside their bodies followed by slurping the nutritious concoction through their mouths, allowing the food to be further broken down inside the body. They perform these distinct phases of eating by having two stomachs called the cardiac stomach and the pyloric stomach. The cardiac stomach is a sac-like organ located centrally from where the arms radiate out. Its feeding begins with wrapping the arms around the prey to immobilise it. After seizing the victim, it pries open the shells with its hundreds of suction-cupped tube feet. It is able to squeeze between tiny cracks, making this predator most efficient at opening a variety of molluscs and other types of shellfish. The cardiac stomach is extended out to completely surround its prey before an array of digestive enzymes are secreted to initially commence the breakdown of food. This allows the partially digested food to be retracted and passed into the pyloric stomach, where digestion is completed. Such adaptation allows starfish to eat animals much larger than its mouth.

Like all species of starfish, the crown-of-thorns starfish, *Acanthaster planci*, feed often, and their size depends on the amount of food they eat, not on their age. They are known to cause catastrophic damage to the marine environment from such predatory and gluttonous feeding behaviour. Massive outbreaks of crown-of-thorns starfish comprising tens of thousands of individuals have been known to cause serious harm to the productive coral reefs in the tropics.

Starfish feed often, and their size depends on the amount of food they eat, not on their age. A single starfish has been shown to devour over fifty young clams in a week. Depending on weather conditions and abundance of food in the sea, the population of starfish can explode, resulting in their consuming the entire beds of shellfish on the sea floor. A well-known carnage caused by such predatory and gluttonous feeding behaviour of starfish is the outbreak of a starfish known as the crown-of-thorns starfish, *Acanthaster planci*. This culprit is a well-known coral predator. Massive outbreaks comprising tens of thousands of individuals have been known to cause serious harm to coral reefs in the tropics. I have seen almost the entire coral reefs fringing Pulau Mamutik off Kota Kinabalu, Sabah, devastated by the crown-of-thorns starfish in 1986. Both hard and soft coral were totally devoured, leaving the reefs whitish in colour instead of the usual iridescent hues and colours. The once healthy reefs were destroyed within days. Some believe such an outbreak was a natural phenomenon. The unusually warm weather had resulted in excessive availability of food and nutrients, which drove the crown-of-thorns starfish to indulge in their gluttonous feeding behaviour. This increased their reproductive rates and growth explosion of the starfish. But some marine ecologists believe the phenomenon is due to the depletion of the giant tritons, *Charonia tritonis*, the primary predator of crown-of-thorns starfish. In developing countries like Malaysia, Indonesia, and the Philippines, giant tritons are collected for its large attractive shells, which are sold as tourist souvenirs. Over-collection of this predator could have caused the explosion in the crown-of-thorns starfish's population on the reefs. And because starfish by nature are voracious eaters, they would make a glutton of themselves feeding non-stop on the meaty part of the hard corals despite being enclosed in hard exoskeletons.

The Klias Peninsula in Sabah is like a savannah except it is lush and green with mangroves. A consummate biologist teaching at the Universiti Kebangsaan Malaysia, Sabah Campus (UKMS), in the 1980s, Robert Stuebing, and I used to cruise along the main river feeding the peninsula looking for crocodiles. Rob had a research grant to survey the crocodile population of this area, and I was interested to see if crocodile spotting could be turned into a tourism asset for this sleepy local settlement. Night crocodile spotting had been promoted as a tourism activity to supplement the income of people dwelling near mangroves in some places in Sabah. The local youth had the opportunity of earning extra income from becoming tour guides in this growing nature activity. Rob purchased a small boat for use in the survey, which he safely stored underneath a stilt house belonging to a village imam, a polite elderly religious leader of the community. We only used the small outboard-powered boat on weekends for crocodile surveys—a convenient arrangement instead of towing it all the way from the campus.

I still remember a particular occasion just before setting out on one of our survey trips when the curious imam asked Rob what he was looking for in the vast mangroves across the river in front of his house. Nonchalantly, Rob stated his intention in almost perfect local dialect, 'Cari buaya'—meaning, 'Looking for crocodiles'. The imam wasn't impressed or worried. He blankly looked at us and in a cynical tone said, 'Son, I've lived here for more than thirty years and have been up and down this river fishing hundreds of times. I've yet to see a single crocodile.' He sounded very serious and wished us good luck. We nodded with a smile and went on our way.

We were playing for time waiting for the sun to go down and darkness fall before the eye-spotting began. I couldn't think of anywhere in the mangroves more peaceful than on a small boat at a standstill in mid-river. The chorus of insects surrounded us as we took bites of a variety of authentic *kuih-muih*, the traditional sweets bought from the roadside stalls on our journey just minutes prior. The ever familiar pandanus aroma and taste of palm-sugared glutinous rice simply stole our hearts. Against a blazing sunset panorama at a distant, it was the best place to be. Even the mosquitoes left us alone to enjoy our local delicacies on the calm surface of the ebbing river. Perhaps the pesky mosquitoes couldn't be bothered to fly far for a blood meal from us sitting way out in the middle of reasonably wide Klias River. At the same time, it took some courage to pay no heed to the fact that in the water around us were also feeding grounds for estuarine crocodiles twice as long as our drab little boat. But to discount and feel comfortable with that must have come with Rob's familiarity with the locations and time of the day we were doing our survey.

With a battery-powered torchlight on his forehead, Rob scanned the surface of the water about thirty metres ahead of our boat. With bated breath, we were absolutely focused on seeing two red reflective spots on the water. It didn't take much field experience to see twinkles of red in the darkness of mangroves using this technique. But it took some learned familiarity to determine if these spots were eyes of crocodiles or those of other creatures such as frogs or spiders. It was easier to spot little baby crocodiles of not more than forty centimetres long cruising on the water surface. Juveniles were much less suspicious of us than the adults. If we were to approach the little ones, they would calmly float away from our boat till they eventually decided to sink and hide amongst the mangrove roots. We could have been too close for comfort in their little minds. Despite their fear-provoking reputation and enormous size, adult crocodiles didn't seem unduly circumspect of our presence on the boat. More often than not, we would have a brief glimpse only of their eyes and nostrils before they disappeared beneath the surface of the water and swam with barely a ripple to give away their location. Most adult crocodiles were secretive and surprisingly devious in avoiding us. That particular night, we came across one rather clumsy adult crocodile of about two metres in length. It was resting in the water near a tree. It seemed to be unperturbed by our approaching boat. But suddenly the poor beast panicked, and with a single whip of its tail, it lunged sideways and collided hard, snout first, into a huge log. It looked stunned for a moment, probably quite embarrassed, before collecting its senses and sinking gracefully below the surface and swimming away. That night alone, we spotted a dozen crocodiles in the water just a few hundred metres from the riverbank where the village imam's house was sitting. He'd be shaking his head in disbelief if we were to tell him of our many sightings that night. Understandably, our imam friend was never out there in the darkness of night specifically looking for crocodiles. Even if he were, he wouldn't be using the spotting techniques Rob was rather skilful and competent at.

In the wild, crocodiles eat insects, fish, frogs, lizards, crustaceans, and mammals. They are far from selective eaters. It is often said a crocodile feeds on anything it can outswim or ambush and overpower. In some places of Borneo, crocodile attacks on humans are common in habitats where large crocodiles thrive close to human settlements. Some rivers within the Sarawak mangrove areas along Batang Lupar are famous for fatal crocodile attacks on humans. About eight out of ten crocodile attacks on humans in Malaysia occur along major rivers of Sarawak.

Naturally, most people are puzzled by the feeding habit of crocodiles. They are known to save their prey at the bottom of the river until it begins to rot before devouring the meal. There could be several reasons why they do this. Crocodiles do not eat as often as other animals; they only eat about once a week, around fifty times a year. In extreme cases, crocodiles have been documented as existing for a year without eating. Storing food is simply the crocodile's way of stocking the larder for times of food scarcity. They have an extremely robust digestive system, so rotting food is not going to give them food poisoning or gut aches. Their digestive enzymes and acids are more than a match for any bacteria found on carcasses. The most plausible reason for putting their food aside to rot prior to eating lies in their teeth. Their dental structures are not designed for chewing but more suitable for severing and tearing meat apart. This is why crocodiles habitually gulp down their meal in huge chunks. They are not able to break food down into smaller pieces with their teeth. This also means they cannot derive the most effective energy return from fresh meat. Leaving the meat to rot is essentially a means to save energy from having to chew hard on the tough fresh meaty food. It is a pre-digestion mechanism. The rotting meat is softer and breaks apart easily, allowing their digestive tract to expend less energy. An analogy is often drawn with an infant fed on a purée instead of chunks of fruits or vegetables. Digestion uses energy, and the less energy used for digesting food and absorbing nutrients, the net benefit is greater. Like a growing infant, a crocodile receives the maximum energy output with the minimum energy input.

Crocodiles are known to store their prey underwater to rot before devouring. Habitually, they only eat about once a week, around fifty times a year. But the most likely reason for putting their food aside to rot lies in its dental structures, which are not designed for chewing. Instead, their teeth are more suitable for severing and tearing meat apart and not for biting into small pieces as occurs with other carnivores. This is also why crocodiles habitually swallow their meals in large pieces.

There was a notorious man-eating crocodile ironically known as *bujang senang*, or 'happy bachelor', of Batang Lupar. Since the early 1980s, this estimated four-metre-long beast struck terror amongst the people living along the river as human toll continued year after year. It managed to avoid being hunted and exterminated by professional crocodile hunters for many years. In 1986, Rob and I volunteered to give a helping hand in luring the beast by using recorded stress calls of young baby crocodiles. We caught a baby crocodile at a crocodile farm in Kuching and beat its tail with a heavy rod, making it cry under pain and duress. We recorded the stress sounds and played them in the still of the night from a boat where the 'happy bachelor' was supposed to be lurking for its next victim. It was an interesting technique to lure the mother crocodile to our boat out of curiosity. We managed to attract the interest of a mother crocodile one particular night. The mission was to kill the man killer from the boat. In our team were two trigger-happy sharpshooters from the police department ready to execute the operation. A rough estimate of the size of the mother that responded to the stress calls was quickly derived using a formula by plucking in the perceived distance between the two red eyes we saw from the boat. It was too small to qualify as the man-eater *Bujang Senang*. Both Rob and I were quite relieved our two trigger-happy hunters didn't get to fire any shots. Nothing came close to completing the mission at the end of the night.

Along Batang Lupar, locals don't go swimming in the river as they used to do for recreational activities. In the past, they would float on tyre tubes down the river on weekends and wash clothes and kitchen utensils in the river. On fishing trips, they would unhesitatingly jump overboard to retrieve entangled fishing lines or lost lures. They do not do that nowadays because they are not sure what is under the water surface. The crocodiles are becoming too close for comfort even on land. Villagers often find crocodile tracks on the way to their inland farms. It may be the path of a mother crocodile on her way to the nest of babies in the grassy patches on the open land. People are justifiably petrified at the thought of confronting a crocodile mother defending her babies while on the way to their farms. Their lives, especially the free-roaming children, are at risk every time they venture close to nature in the water or on land. There are just too many crocodiles around today. Many people want them culled for their safety. Today in Sarawak, crocodiles are protected under the Wildlife Protection Ordinance 1998. Since crocodile hunting is outlawed, the sightings of monstrous crocodiles sunbathing along Sarawak rivers have become more common.

But mixed opinions abound on calls for the culling of crocodiles. Some believe the protection of crocodiles has become extreme and their numbers need to be controlled. Some cheekily argued that if we don't run around with dinosaurs on land, why should we swim with them? The villagers consider their views have been ignored, stating that the person who doesn't support a cull fails to see the danger for and safety of their children. One needs to have lived beside this river for more than thirty years to realise the population of crocodiles has indeed increased to dangerous numbers. There are outspoken supporters of not culling as well. Despite data showing unprecedented growth in crocodile numbers over the past thirty years, conservationists and scientists are not in favour of culling crocodiles. Their compelling argument is that removing some crocodiles would just cause a vacuum in the habitat that would only result in other individuals moving in to occupy that void. So, do we just keep shooting them until we get an overall reduction in numbers? Culling would not really achieve anything. At the end of the day, a menacing animal like *Bujang Senang* is just another animal we need to treat as yet another natural instinct and behaviour to live with. Personally, I just love the mangroves not just for their natural beauty but also for their unique biodiversity. They are pregnant with natural wonders and fascination. I'm a great believer in our need to learn more about the iconic species thriving there, and it is our responsibility to ensure their continued survival by sharing this unique ecosystem with them. For the people living in the vicinities of this magical ecosystem, opportunities abound for ecotourism potential including crocodile spotting, watching the proboscis monkeys, and marvelling at the spectacular light display of the fireflies.

Unlike plants or algae, animals cannot make food from water and carbon dioxide by photosynthesis. Animals, including humans, need food from external sources. For growth, they need to sustain many vital metabolic processes including respiration, digestion, transportation, excretion, circulation of blood, and reproduction. They must eat foods rich in the essential elements of proteins, carbohydrates, fats minerals, and other nutrients. For humans, there's another dimension to eating. Today, eating isn't about becoming full and obtaining sufficient energy to carry out our metabolic processes and daily activities. We are also conscious about eating healthily. Watching what we eat is fundamental to our good health and well-being. We eat to maintain a healthy weight and to reduce our risks of metabolic disorders like diabetes, high blood pressure, high cholesterol, cardiovascular disease, and some cancers. Humans have larger brains to invent tools with which to grow food. Contrast that to animals. Animals have to seek far and wide for natural food and often have to compete with others to access it. To survive, they have evolved ingenious means of obtaining food. They consistently work hard to get enough food to eat. The behaviour and competence of mangrove animals in finding food are various and perplexing. In their own way, the species possess ingenious means whether amongst those flying high in the air, within the mangrove plants, on the mudflats, on the beach, or dwelling in the depths of bodies of water. Animals do not seem conscious and preoccupied with what they eat as long as food is edible in taste and texture. For animals, eating is what they do to obtain sufficient energy to perform life processes to survive.

11

Making More Of The Same

The whole purpose behind evolution, adaptation, and natural selection is to have robust clones with traits that are beneficial to the organism, resulting in biological features and metabolic processes that can be passed down, allowing them to adapt to the environment better than other organisms of the same species. In the end, their survival and reproduction are better compared with other members of the same species or those around them. Inheriting these superior traits is crucial for the continuance of the species, which is achieved through reproduction amongst their own species. The means of producing more of their same kind amongst mangrove species are varied and most fascinating. The rituals and energy expended, which are adopted when finding mates, are uniquely tailored to reap success against all the odds thrown at them by the harsh mangrove environment.

The fundamental purpose of every animal's life is therefore to produce more of its own kind. No species leaves this ultimate purpose to chance, and evolution has been giving a helping hand to all life on Earth. Neither do humans, for all intents and purposes, deviate from this purpose. We live, knowingly or not, to produce more of our own kind. We, too, work to seek mates we fancy. We dress to impress. In animals too, looking attractive, putting on elaborate rituals of singing, dancing, and displaying physical fitness become important facets of luring partners. During the breeding season, all this can come into play. The males start displaying bold colours to attract the attention of the females. Males with the most colourful appearance have the best chance of finding mates to produce more and superior offspring. Others use sound. Hence, birds and insects employ coded calls in their chirping and stridulating. Some even complement their singing with dancing to attract mates. Others release special scents only to be recognised by the same species. Interestingly, some resort to offering gifts as humans do. Many mangrove species evolved characteristic signals to increase their chances of attracting mates to reproduce. Each species has their own way of locating and luring mates, which is successful for them.

Weston, Sabah, is a small town located on the bank of the Weston River, which feeds into a major part of the Klias Peninsula wetland. It has become a popular destination for viewing the iconic proboscis monkeys as visitors need to drive 125 kilometres only from Kota Kinabalu to Weston. As well as the enthralling river cruise, Weston also offers a glimpse of its historical past during the British colonial era. Weston was named after a railway engineer, Arthur J. West, in 1889. The once buzzing little town was a principal meeting point for trains and ships because of its strategic location at that time. The row of old wooden shophouses built during the British colonial period over sixty years ago is still standing and in use. A two-storey building nearby is a Chinese primary school built in 1932, probably one of the oldest wooden schools in Borneo. The old timber jetty, which has existed since 1896, is another legacy of the British rule. Weston port was abandoned when the British North Borneo Company learned that the shallow water of Weston was no longer suitable to serve as a deep-sea wharf. Today, Weston is a tranquil water village of not more than two thousand people. The majority of people here dwell in stilt houses above water linked by a network of wooden walkways. Because of its fame as a seafood eating destination, local businesses are still thriving to complement the growth of the ecotourism industry.

Just as darkness fell, we took a small boat of about eight-person capacity to cruise along the river towards the mangrove forest nearby. Quite a contrast to the hot humid ambience of daytime when your eyes can scan the green landscape as far as your eyes could see, mangroves by night seemed pitch dark and relatively spooky. The bewildering maze of narrow channels between the dense tangle of exposed mangrove roots contributed to such an unnerving eerie atmosphere. It was near low tide, and I could see the water ebbing when I scanned the muddy riverbank with my torch. The air hanging over everything smelled of brackish water and black mud. Echoing through the dark mangrove forest were the bizarre gunshot-like sounds of the mangrove snapping shrimps in response to the changing tide. A host of insects of different sizes zigzagged in the air space above our boat, too fleeting to even provide a clue as to what insects they were. But the bats seemed untroubled as they provided before us a dazzling acrobatic air display above the water surfaces. They were actively skimming, swooping, and snatching insects for food.

Then as the boat took a sharp bend of the river, it suddenly felt as if we had just arrived in Disney Wonderland. Christmastime seems all year round in Weston! The riverbanks were sparkling with tiny yellow lights. Constellations of these glittering lights seemed to blink in synchrony within the foliage of the trees that line the mangrove edges. They conjured images some might see as romantic, mystic, mysterious, intriguing, or even haunting. We were slowly cruising along the river to witness this famous 'Christmas lights display' along the tropical mangroves, as aptly claimed in tourist brochures and on internet sites. My squinting eyes followed the path of a few small yellow lights flickering at a distance, which gradually floated away into the darkness. They were colonies of congregating fireflies known as *kunang-kunang* in Malay. In the world, there are over two thousand species of this soft-bodied beetle of five to twenty-five millimetres in length. The congregating firefly species from the genus *Pteroptyx* found in Sabah is the most impressive and famous for their habit of gathering and flashing in numbers ranging from hundreds to thousands of individuals per group. They create one of the most remarkable natural wonders to experience in absolute delight and awe.

Biofluorescence in fireflies is more than just sparkly magic on a humid tropical night of the mangroves. They are not there to dazzle man's visual senses. Those flashes are purposive and highly crucial in ensuring their genes are passed on to the next generation. They glow and flash to communicate, mate, and reproduce for the continuance of their species. Fireflies seem to spend their adult lives courting and mating. They're very single-minded in this sense. A billion bursts of light are produced to attract mating partners. For this, they have evolved to have a special light organ like a lantern in their abdomen. Males have two segments of the abdomen that light up, while females have just part of one.

To attract mating partners, fireflies have evolved a special organ capable of producing flashes of light recognised by their own species. Males have two segments of the abdomen that light up, while females have just a small section of one.

I have been fascinated with fireflies since a child growing up in Jelawat, a small village in Kelantan, Peninsular Malaysia. There was nothing more magical than watching these little 'moving stars' bopping up and down against the pitch-black sky. They were dancing around above my head, and village children had a great amount of fun trying to catch these graceful flying 'stars'. My grandfather would assist me in swiping these 'stars' by swinging his *semutal*, a traditional cloth headgear worn by the Kelantanese men. We would be swiping them down to the ground, place them in our palms, and gently keep them in an empty clear jar. We would watch in admiration how the jar would increasingly light up as we filled it with more and more fireflies only to set them free before going to bed. My grandfather would say I'd be assured of sweet dreams free of bad nightmares.

Enchanting flashes of light emitted by fireflies are their language of love used to signal and attract mating partners of the same species. Other firefly species may mimic the twinkling signals to trick fireflies of different species to come closer, which would become food for other species instead of being mating partners.

The aspect that makes fireflies so enchanting is their ability to produce flashing lights like that of a Christmas tree along the mangrove edges, branch to branch, one *Sonneratia* tree after another as the luminosity multiplied. The flickers appeared haphazard and uncoordinated at first but progressively became synchronised. In a matter of minutes, the entire row of trees lining the riverbank was pulsating on and off. It was simply magical. A sight to die for!

This flicking light is a means of communication—a form of insect Morse code. Female fireflies would be repeatedly emitting light in a specific frequency pattern. There are a total of eight species of *Pteroptyx* in Malaysia. Five of them are found in Sabah—namely, *P. tener, P. gelasina, P. similis, P. malaccae,* and *P. valida*. All of them actually flash actively, not just passively glow in the dark. Interestingly, each species flashes at different pulse patterns that are specific for the species. They are not at random intervals as seen by our naked eyes. However, the specific frequency pattern of pulsating light can be recognised by the species producing it. This is how they communicate to attract the same species to mate. One person on the boat aptly commented from within the darkness of the mangroves, 'Fireflies use a torch to find their mating partners!'

As researchers learn more from deciphering the mating codes of fireflies, more similarities are found to animals such as frogs and insects in vocalising their sexual pleas. Females in many groups of animals seem to prefer higher-energy courtship. In some frogs, crickets, and katydids, females appear to become more attracted by longer, louder, or faster calls. This can mean that a female prefers to mate with a courting male who works sufficiently hard to get her attention. Such devotion can be seen as possessing good genes and potentially able to provide for her better. *It is no wonder chatty guys with the gift of the gab seem to attract more females than timid and bashful guys*, the thought crossed my mind. The communication continues until they reach each other and mate. The brief courtship solely relies on each other's ability to recognise the pulses of light signals that are characteristic of their species. There is no involvement of other senses such as smell, touch, or hearing.

Studying biochemistry later in life, my fascination with fireflies grew. I learned about the chemical reaction inside their bodies that enables them to light up. The method by which fireflies produce light is perhaps the best example of bioluminescence. The mechanism has been known at the molecular level. A firefly switches its light on when two proteins, luciferin and luciferase, react inside the abdomen in the presence of oxygen. When oxygen combines with calcium, adenosine triphosphate (ATP), and the chemical luciferin in the presence of enzyme luciferase, light is produced. Unlike a light bulb, which produces a lot of heat in addition to light, a firefly's light is 'cold light' without a lot of energy being lost as heat. This is necessary because if a firefly's light-producing organ became as hot as a light bulb, the firefly would not survive the experience. Interesting as well is the ability of the insect to control the beginning and end of this chemical reaction. Thus, it can start and stop its light emission at its whims and fancies. To produce light, it adds more oxygen to the chemicals in the light organ, lighting up the abdomen, and when oxygen is used up, the light goes out. It is still a scientific mystery how insects like fireflies can transport oxygen to the light organ so quickly without having a complex series of tubes known as tracheoles as in human lungs. Recently, scientists may have solved this intriguing phenomenon. It is the gas nitric oxide that binds to the mitochondria, allowing oxygen to flow into the light organ, where it combines with the other chemicals needed to produce the bioluminescent reaction. Because nitric oxide breaks down very quickly as soon as the chemical is no longer being produced, the oxygen molecules become dissociated and recaptured by the mitochondria and made available for the production of light. When the firefly light is off, no nitric oxide is being produced. A funny thought crossed my mind on the role of nitric acid as a 'love gas'. Nitric oxide is the same gas produced from taking Viagra, the drug used for treatment of erectile dysfunction syndrome in man. *Is it just a coincidence that nitric oxide has a role to play in the sex life of both fireflies and man?* I wonder.

The ability to create and control bioluminescent signals at will can have its downside as well. It can be manipulated and exploited for dark purposes. In some firefly species, this intricate signalling can spell trouble if known to others. Females of the firefly genus *Photuris* have an appetite for other fireflies. A hungry *Photuris* firefly that learned to mimic the flash patterns of other species can use the flash patterns to obtain food. Male fireflies from the *Photinus* or *Pyractonema* genus often fall for this trickery. From below, *Photuris* fireflies will emit flashes in sequences and intensity similar to those produced by *Photonus* and *Pyrotonema*. Seeing and recognising these fake 'Morse codes' as those from their own females, the males will be tricked to fly closer until it is too late, ending up not as a mate but the next meal for another species. It is not clear if such risky love affairs also occur in the *Pteroptyx* genus of Sabah fireflies.

Clearly, fireflies have developed an ingenious method of attracting the attention of potential mates in the darkness of night—a challenging one that has succeeded very well for the continuance of firefly's species. How about getting noticed in a vast open space like the mangrove mudflat? That, too, can be challenging for little creatures crawling about trying to find mating partners. Many animals inhabiting the mudflats like bivalves and crabs have evolved and adapted mating behaviours unique to their groups.

Female and male fiddler crabs have claws that are starkly different in size. Females have two small claws they use to pick up bits of sediments on the mudflat from which they extract microscopic food. Adult males, however, have one small claw and another that is disproportionately larger, constituting up to half its total body weight. This greatly enlarged claw has a crucial role in the mating behaviour of fiddler crabs. It is of the essence in reproduction and continuance of the species. When courting females, male fiddler crabs resort to waving their huge claw in the air to attract attention. Such attention-grabbing gestures are visible, unobstructed to potential mates even from a distance on the vast open mudflat. The object is to lure the female partner to his burrow. Once a female crab approaches, the male becomes more animated using both gestures and sound to further impress. He will perform bizarre rituals at the entrance of his burrow, which can be quite elaborate. Prior to displaying luring and seductive gestures, the fiddler crab meticulously builds a sand 'hood' at the entrance to his burrow, probably to motivate females to accept his humble abode in case his beautiful claw isn't acceptable. Dancing is a part of courtship behaviour. The male fiddler crab darts back and forth, waving his large colourful claw in mid-air. The female spends some time just watching and seemingly uninterested in the performance initially. But the male relentlessly repeats a similar pattern of choreography in his attempt to impress her. This might include stepping back and forth, jolting, or jumping slightly while waving his claw side to side. He also uses auditory signals. To create sound, the male crab strikes the substrate with the lower base of its large claw, drums on the substrate with both claws, and taps the ground with its walking legs. Such energetic calling bouts can last several minutes. Waving his major claw rhythmically almost non-stop until adoringly taken on by the willing female is surely a task requiring a lot of energy. Further, while busy with these soliciting rituals, he at the same time keeps a close watch on other males nearby threatening to steal his potential 'bride'.

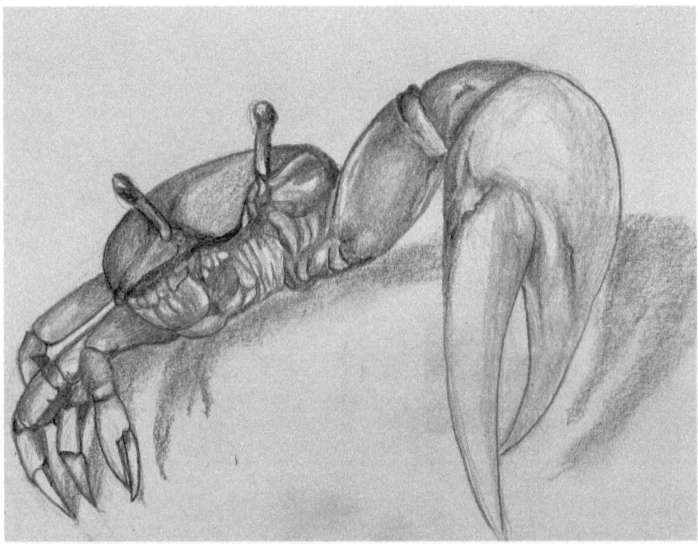

Adult male fiddler crabs are endowed with one small claw and another that is disproportionately larger, constituting up to half its total body weight. This greatly enlarged claw has a crucial role in the mating behaviour of fiddler crabs.

The fiddler male is endowed with not only a long attractive claw but also an effective indestructible weapon during fights. He uses his large claw to fight other males over burrows. The size of the muscle attached to this claw is testament to its role in protection by applying forceful pinches on prey, predators, or competitors. The hinge of the claw is capable of the strongest grip. As the claw grows long and more attractive to females, the cutting edge of this needle-nose pliers, too, becomes sharper. Millions of years of evolution have created a male fiddler crab that is not only well sought-after for mating but also very powerful to ward off competitors or predators.

To motivate the mate, male fiddler crabs resort to displaying luring and seductive gestures, which include energetically waving his big colourful claw in the air at the entrance of his newly built sand 'hood' to his burrow.

Clearly producing more of your own kind is a means not without competition in the mangroves. Beauty is certainly in the eyes of the beholder in the world of fiddler crabs, where huge strikingly colourful claws become the primary sex appeal to mates. The male proboscis monkeys do not have to display shimmering fur, sing, dance, and strut around to impress mates like fiddler crabs. Female proboscis monkeys are only interested to mate with males having the biggest and longest nose. The size of his nose seems the only criterion. The object in producing more of the same kind here is to inherit the trait of having an equally big nose if not bigger.

Males compete amongst themselves for females' responsiveness to copulate, and only the one with the largest nose wins. After generations of selective breeding, the male noses have become absurdly large and relevant to the reproduction of the proboscis monkey. In the world of the proboscis monkeys, nose size matters. The larger, the better. As a reward, the successful male will acquire a harem of up to a dozen females. This is evident from the larger number of females in the harem of big-nosed males compared to that belonging to males with a smaller nose.

Sexual preferences in humans are but a private matter and usually best kept between a person and their partner. For most animals, however, evolution has made their sexual preferences an open secret. A big, long nose is considered sexy in the world of proboscis monkeys, giving males with the longest nose a better chance of attracting mating partners.

Proboscis monkeys have a well-defined social structure. They are always found in two types of groups: the bachelor group and the one-male harem. Bachelor groups will consist of five to ten male monkeys with no dominant leader. The harem consists of one dominant male and ten to fifteen females. As head of the family, the male becomes the master of all the females and their babies. While heading the group, the dominant male proboscis monkey seems preoccupied with sex. He will try to father as many offspring as soon as possible. He is in a hurry to produce many offspring as quickly as possible because he has to procreate before his time is up. Time will soon come when he will be displaced by younger mature males from the bachelor groups. When this happens, he will be banished from his harem, and all dependent infants in the harem will be killed by the victorious male. This malicious and spiteful act is undertaken by the new dominant male to get the female monkeys pregnant with his own offspring as quickly as possible. To signal that he is ever ready to copulate, a male proboscis monkey sits with his legs akimbo, fully displaying his perennially erect genitals. This is his own way to get noticed by his many 'wives' in the harem. Females, in contrast, sit with their legs closed, prim and proper as ladies. Studies have shown that copulations in this monkey species are in fact initiated by females. She chooses the male to mate with and signals her choice by displaying in front of him.

A male proboscis monkey sits with his legs akimbo, fully displaying his perennially erect genitals. Perhaps this is his own way to be noticed by his many 'wives' for his ever-ready desire for mating. Females, in contrast, sit with their legs closed, prim and proper as ladies.

Generally, birds sing to attract mates. Rendering these elaborate songs is something they learn early in life. In the breeding season, the male birds sing their unique songs, which can be recognised by females of their own species. They also sing when defending their territory. Technically speaking, the sounds that birds produce can be categorised into two distinct classes: calls and songs. A call is usually a short and simple vocalisation that signals flight or danger. A song, however, tends to be a long and complex vocalisation produced during the breeding season. The arrangement of the high and low notes and the pauses between each segment of the song is important. They must be sung in correct sequence. Usually, songs are organised into several phrases or motifs consisting of series of syllables, which, in turn, are made up of a collection of single notes or elements. Each bird produces its own repertoire of songs akin to each human singer performing their different versions of the same song. So even between species, there exist many variations in the repertoire size or song type. The number of songs in each repertoire can range from single to more than five depending on the bird species. In many species, only males sing; while in others, both males and females sing equally in duets.

Bird songs are generally for attracting mates during the breeding season. Other courting or mating behaviour, however, includes the practice of rewarding offerings to their mates, just as humans give presents to spouses and partners.

The most remarkable observation about mating behaviour of birds is the practice of rewarding or giving offerings to their mates, just as humans give presents to spouses and partners. Many male birds bring food to the females as evidence of their ability to care and provide food for their future chicks. This also serves to convince the female she will receive nourishment while she is producing and incubating eggs. The male kingfisher has been documented to use gifts to charm its partner. He presents his gift, usually morsels like a small fish, grasshopper, or worm, to the female bird. If the gift is not acceptable to the courting partner, he eats it himself and flies away to find more appetising and acceptable food. He may have to repeat this same bribing performance to persuade the female to accept his offer. In full view of the female, he may also entice her by demonstrating his fishing skill by diving head first into the water to spear a fish with his dagger-like bill. Successful, he flies back to feed his prized catch to the waiting female.

The male kingfisher uses gifts to charm its partner. He presents his gift, usually morsels like a small fish, grasshopper, or worm, to the female bird till he gets the approval and appreciation of the gift by his courting partner.

The idea of 'corrupting' females is present in spiders too. The much smaller male spider is up against some real odds. He needs to be cautious when approaching his potential partner, which is by nature more aggressive and larger. Charm is required. The male orb-weaver spider strums a unique signal on the web to catch the attention of his mate. He produces special vibrating frequencies from the edge of the female's web to make sure she does not mistake him for prey and becomes a meal instead of a mating partner. Interestingly, the male web spider presents the female with a bridal meal. The gift comes in the form of dead insects wrapped in silk. It will be presented to the female just before the couple copulates while the female dines on the gift. This is, however, carried out amidst cautiousness and vigilance all around. The timing for the presentation of this gift is absolutely crucial to avoid the female making a meal of the male as well.

After all the highly sophisticated and romantic ways animals are known to employ in trying to gain the attention of a mate, the last animal one would expect to find a similar approach is with the crocodiles. Huge and heavy, one would think crocodiles would be the most unlikely creature to perform courtship rituals, let alone singing and dancing. How wrong can we be? In the build-up to the breeding season, male crocodiles are known to stimulate courtship and mating behaviour by advertising themselves in conspicuous displays. Often, they serve to either drive off rival males or attract receptive females. They may not break into fanciful energetic dancing to impress females, but they are known to produce a hardly audible rumbling sound before bellowing loudly, sending vibrations to the water surface. The water next to it would be shimmering and bubbling, making the male more noticeable. The vibrations from this sound can be detected by females from a distance. Courtship observed in captive crocodiles involves snout contact and rubbing, body riding, vocalising, bubbling, and circling, with both partners constantly submerging and resurfacing. In the wild, fights between rival males do occur occasionally prior to copulation. This includes chasing, head slapping, and growling, which can escalate to full-blown combat, leading to serious injury and even death. Females also become intolerant of other females and will jostle for dominance. The victorious pair will indulge in copulation, lasting up to fifteen minutes, which may take place while totally submerged. Defeated rivals within the vicinity also heed the warning and move away peacefully.

Despite their fearsome appearance, estuarine crocodiles are known to be devoted mothers. Female crocodiles pile mud, sticks, and plant materials to build a nest before laying about twenty to sixty eggs. The female periodically visits the nest during the eighty-ninety-day incubation period. When hatched, female crocodiles excavate the nests to free the hatchlings. They are gentle and cautious too. They have been seen gently biting the eggshell as the new hatchling is struggling to hatch. Sometimes, female crocodiles carry their babies inside their gaping jaws, transferring them from the nests in the hot open sun to cooler sheltered areas under shrubby vegetation. The young ones stay with their mother for the first two years of their life.

Despite their aggressive reputation, estuarine crocodiles are devoted mothers known to be gentle and cautious towards their babies. They have been seen gently biting the eggshell as the new hatchling is struggling to hatch and would sometimes carry as many as fifteen babies inside their gaping jaws, transferring them from the nests in the hot open sun to cooler sheltered areas under shrubby vegetation. Mother's instinct prevents her from closing her jaws, ensuring the young ones are safe from her large, sharp teeth.

Obviously from the discussion above, the harsh, inhospitable mangrove ecosystem has failed to drive and wipe out species to extinction. In fact, this unique tropical ecosystem serves as a suitable platform for the reproductions of species that are unique to the environment. Mangroves have provided myriads of selective pressures to create a range of robust clones with biological adaptations that have continued to exist till today.

12

Game Changer In
Climate Change

Over the years, I finally realised I have moved away from my own scientific specialty and ventured into another professional domain. Despite having trained as a microbiologist and immunologist, I have attempted forays into areas in which I wasn't totally comfortable. I was driven by my naive childhood fascination instead. I quickly became committed and passionate about the environment and conservation issues. However, I was relieved to have reassured myself that to stray from my own area of expertise was also out of necessity for the benefit of my country. Malaysia was in a fragile and precarious state throughout my years of service as an academician from the 1970s. The state of its environment and associated issues were only a cursory agenda put on the back burner. That hadn't changed till I retired a few years ago. Through the years, I bore witness to the rapid growth of economic progress and at the same time was worried about the many adverse impacts it brought to the environment. The conservation and sustainability of its natural resources have not been this vital since the country's independence in 1957. The runaway train headed for disaster needs to be stopped. Taking some of the issues in my hands, I did a few things to make a difference. I decided to focus my attention on raising awareness in a few peripheral and fuzzy aspects related to environment and conservation. I chose to do that through education. I embarked on a theme that revolved around awareness of nature. Indeed, it was a deliberate choice to make my mission sound more ambiguous than ambitious. This was because my own mind seemed still shrouded with vagaries and uncertainties about the effectiveness of my mission using an educational approach. The definition of nature itself was confusing. Does nature collectively include the physical world comprising plants, animals, and other features of the earth? Nature is essentially everything but humans and human creations. More debatable is the importance of nature to humans. The reluctance to admit so is deafening. Yet more and more we see nature underpinning human advancement in our progress to become sustainable in our existence on planet Earth. It's escapable that humanity must rely on and learn from Mother Nature. Our forests, rivers, oceans, and soils provide us with the food we eat, the air we breathe, and the water we need to keep us alive. By deliberately absconding the generous benefits from nature's bounty, we are inviting huge social and economic costs for ourselves. My intention was very serious about these matters. But I have refrained from going too deeply into these issues at an academic level. Instead, I consistently stayed close to the surface of things, mostly through education. I found myself participating in seminars and conferences on issues relating to environment and conservation. I was equally busy with writing popular articles for non-academic magazines, targeting laypeople as my audience. One of my favourite topics has been the wrath of abuses causing the mangroves to be in much peril.

We are increasingly recognising the many roles mangroves play to provide shelter for fish, to protect the coast, and to keep coral reefs healthy. Mangroves have been providing a slew of benefits for us with their strategic locations lining our coasts. Mangroves have protected us from sudden storms, surging waves, and deadly tsunamis. They have served as barriers to reduce gradual erosion of our shorelines. Hundreds of marine species, which provide food for us and the ecological balance, use mangroves as shelter and nursery for their young. Local villagers depend on mangroves for their livelihood. Here, they catch a variety of valuable seafood for their own consumption and for commercial purposes. Mangroves are a supply source for wood as building materials and for charcoal. Unique endearing animal species living in mangrove forests are natural assets bringing in wealth for the ecotourism industry. It's easy to see money from mangrove protection. Tourism operators and commercial fishers all have an interest in keeping mangrove coral communities healthy and in abundant supply. Even the oft-ignored local communities living in proximity to mangroves cannot earn a living in the absence of this productive ecosystem. All life on planet Earth are clear beneficiaries. This is, however, true only if these wetland ecosystems remain intact and preserved till the end of time. Reasoning from the economic perspective has been the preferred way in defending the mangroves. The valuation on mangroves in dollars and cents only becomes attractive when their biological services are duly taken into account. Indeed, if we aim at raising awareness and gaining acceptance of stakeholders, the economic angle of environmental accounting should be adopted. Researchers estimate the monetary value of the benefits from ecosystem services provided by mangroves at $194,000 per hectare annually. Multiplied by their global extent, that means the world's remaining mangroves provide around $2.7 trillion in services every year. Recent studies estimate the real value of mangroves could be even higher.

While those economic services are well and good, there is a role that has been played by mangroves for millions of years that we have undervalued knowingly or otherwise. There exists a role mangroves play that we least appreciate. We simply can no longer continue to be unaware and unappreciative about this role—that is, its role in reducing global warming! The role of mangroves in thwarting global warming and climate change needs to be recognised and valued immediately. Failing to do this will lead to catastrophic costs. Wetlands such as peat swamps and mangroves are critical ecosystems in our fight against climate change. However, presently, this role is grossly unappreciated. We can never overemphasise this crucial role that experts are increasingly proclaiming, 'Mangroves is the game changer in our climate change debacle. In this context, it's hard to fathom a more valuable ecosystem than a mangrove forest.'

Effective action on climate change will require a combination of emission reductions and atmospheric carbon removal. The two actions must go hand in hand. At the 2021 United Nations Climate Change Conference COP26 in Glasgow, UK, every participating nation seemed preoccupied with this mitigating measure of reducing emission. Reducing carbon dioxide release to the atmosphere is well and good, but protecting, enhancing, and restoring natural carbon sinks must also become one of our priorities. Mangrove forests can play an important role in carbon removal because they are amongst the most carbon-dense ecosystems in the world. Kept undisturbed, mangrove forest soils can serve as long-term carbon sinks. Wetlands, including the peat swamps, seagrasses, mangroves, and salt marshes sequester far more carbon per year than do tropical rainforests. Of all natural ecosystems, mangroves have been shown to sequester carbon the most. Being primarily in the tropics, mangrove forests are highly productive. Their carbon production rates are equivalent to that of tropical rainforests. They store far more carbon per square mile than do tropical forests. Carbon is sequestered in large reserves in the soil and in the dead roots. Considering the total carbon sequestration by the world's forests, mangroves store about 1 per cent. That may sound small and might be of no consequence in thwarting climate change, but because of their locations along the tropical coasts, mangroves account for as high as 14 per cent of carbon sequestration by the global ocean. This emphasises the potential impacts on the severity of weather patterns and global climate change.

Mangrove forests can play an important role in removing carbon dioxide from the atmosphere and sequestering carbon in their dense root system.

Mangroves are especially suited for carbon capture because they pile most of their carbon on the ocean floor, while terrestrial forests keep most of carbon in trees and branches. Scientists have found significantly more carbon locked up in mangrove soils than previously believed. Mangrove soils held around 6.4 billion metric tons of carbon in 2000. This is highly relevant data to ponder over. Although mangroves are only in tropical areas and cover an estimated 140,000 square kilometres, less than 3 per cent of the extent of the Amazon rainforest, mangroves are powerhouses when it comes to carbon storage. Pound for pound, mangroves can sequester four times more carbon than the rainforests can. Most of the carbon is stored in the soil beneath mangrove trees. If mangrove carbon stocks are disturbed, the resultant gas emissions could be very high. This reinforces the paramount importance of preserving existing mangroves in our attempt to rein in global warming caused by carbon dioxide emissions. Between 2000 and 2015 alone, around 30 to 122 metric million tons of soil carbon was lost over a mere fifteen years. More worrying is when the amount of soil carbon loss is calculated to give the amount of carbon dioxide emission to the atmosphere. It equates to between 111 million and 447 million tons of CO_2, which means mangrove deforestation has released nearly as much carbon dioxide as Brazil did in 2015. Brazil, incidentally, is the world's eleventh-largest emitter of carbon dioxide. This, again, reinforces the immediate need to conserve mangrove forests in mitigating climate change.

Mangroves cover an estimated 140,000 square kilometres in area worldwide, less than 3 per cent the extent of the Amazon rainforest, but they are powerhouses when it comes to carbon storage.

Unfortunately in most developing countries, mangroves are under siege. Research indicates at least 35 per cent of the world's mangrove forests may have been lost between 1980 and 2000, a mere twenty years. Mangroves are disappearing at a rate of 2 per cent per year according to Ecosystem Marketplace's 2014 report. According to the same study, more than 75 per cent of the soil carbon emissions came from mangrove deforestation in just three countries, Indonesia, Malaysia, and Myanmar. The people who depend on mangroves simply don't have the money or political clout to protect the ecosystem pivotal to them. Are developed nations doing enough? The document 'Turning over a New Leaf: State of the Forest Carbon Markets' offered compelling arguments for investing in 'blue carbon' projects. The report emphatically stressed we can slow climate change by saving or restoring seagrasses, tidal salt marshes, and mangroves. The 2014 report was in agreement with the concept that became a point of discussion in 2009 with the publication of two reports: 'Blue Carbon: The Role of Healthy Oceans in Binding Carbon' and 'The Management of Natural Coastal Carbon Sinks'. Unequivocally, all three reports pointed to the importance of simply leaving the sequestered carbon in the wetlands, known as the blue carbon market. In other words, we must invest in the blue carbon by preserving them as they were. It is an intriguingly straightforward proposal for investment towards our future, which is humanity's future. The three reports emphasised that by preserving these wetland 'buffers', we would keep the 'carbon market' alive, allowing it to continue building up what it started millions of years ago.

Mangroves pile most of their carbon in their roots, which end up on the ocean floor, while terrestrial forests keep most of carbon in trees and branches.

Raising awareness on this topic has been daunting. Very rarely do we succeed in our attempt to educate the people in the corridors of power and stakeholders on the need to preserve the mangroves based on climate change validation. We have missed the enormous opportunity to present the narratives through the rationale of reducing emissions. The principle of trying to rein in global warming by reducing carbon emissions can be easily explained and comprehended by most politicians and policymakers. Yet we have achieved little to date. The political will for implementing the Paris Agreement is still lacking despite strident protests by civil society activists and stakeholders. We need to break this deadlock. The climate crisis will become one of the greatest existential threats to humanity. If global warming cannot be limited to a maximum of +2°C above pre-industrial levels by the end of this century, life on planet Earth will be in dire straits. Doing nothing to fix it now will only make the crisis hit humans like a runaway train. Predictions show that continuing with the present emissions trajectory without any mitigation measures will lead to an increase of 3.7°–4.8°C warming with catastrophic impacts. To make matters worse, global warming is intricately linked to climatic issues that could be equally devastating. Human society will be facing more ostentatiously grand challenges, including unprecedented levels of biodiversity loss, land degradation, water scarcity, and rapid urban growth, just to name a few.

Actions on the climate crisis still face political gridlock because of powerful lobby groups shrouded with vested interests and ideological beliefs. The priorities kept losing focus. Countries where governments have close ties to the coal, fossil oil, manufacturing, and agricultural industries show strong resistance in scaling their commitment to reducing carbon emissions. Their agendas of gaining short-term economic benefits are indifferent to climate change. Powerful nations of the world keep postponing the urgently needed low-carbon transformation of our societies, thereby passing the increasingly difficult burden onto future generations. Real or perceived, no other priorities but climate change can lead us towards a sustainable society.

The environmental movements to embark on initiatives to gain visibility and traction on climate change cannot come soon enough. They need to escalate public sensitivities to the many adversities and calamities of global warming now. Climate policies and management must shift to become central and at the heart of public conversation and mainstream media. Strident calls for completing the remaining tasks in operationalising the Paris Agreement and Glasgow COP26 must not fade into oblivion.

It is also time for us to start listening to the science of climate change and global warming. The best available scientific evidence and measures for protecting the environment and biosphere must be adequate and beyond reproach. It is especially daunting to gain public support and engagement on socio-ecological issues. Climate scientists and burgeoning youth-led movements can be of assistance here. Collective effort is key. Youth activists, laypersons, researchers, industry leaders, practitioners of manufacturing and agriculture, policymakers, and politicians need to come together to remove the serious apathy about climate change. Human society needs to understand the impending losses and damages from climate change. The existential risk of sitting back with a wait-and-see attitude is simply too great for humanity.

Earth's average global temperature is now warmer than at any time in the past 125,000 years. The effects are already being felt around the world. As a result of higher temperatures and deforestation, the incidence of droughts has surged. In 2019 alone, tens of thousands of wildfires destroyed 2.24 million acres of natural forests. Humans have eliminated or degraded nearly two-thirds of the world's tropical rainforests according to an analysis by the non-profit Rainforest Foundation Norway. Some researchers worry that rainforests will soon reach a tipping point, a point of no return in which the once-lush forests transition to arid regions. The youth of today are extremely worried about their future as they bear witness to all these early manifestations of climate change. No surprise then that feelings of anxiety, guilt, and grief around the climate and environment issues turn out to be common amongst youth. These powerful emotions can easily spiral into feelings of helplessness and depression. But I am a believer in optimism. This eco-anxiety and worry prevalent amongst today's young people can also be potent motivators for good and healthy coping strategies. They can lead to more thinking, awareness, sensitivities, and communications about climate change.

After I watched Al Gore's movie *An Inconvenient Truth*, climate change became a household subject spoken at dinner tables and in living rooms. I was particularly interested around the issues pertaining to changes in biodiversity. Climate change is already impacting biodiversity. Biological changes are being observed globally. This seemed to cause anxiety and guilt in older individuals like me because I could relate to the ramifications of biodiversity loss in my immediate future. After all, in my lifetime, I have witnessed the extinction of the Sumatran rhinoceros in Sabah. The effect of biodiversity loss was able to prick my conscience. But appreciating the catastrophic effects of global warming seemed harder. The ecological anxiety and guilt seemed more prevalent in young people than in older individuals. Generally, all informed people are worried about climate change, but there is more worry in young people because they picture a considerably dark side of their own global future. It is about their personal futures. In older adults, there seems to be a disconnect between worry about the global future and worry about their own personal future. The majority of teens and young adults said climate change made them feel angry and afraid. Such emotion is somewhat less in older age groups. Having experienced extreme weather conditions and a global pandemic rooted in human destruction of natural forests has made today's young people think about their children and whether they should have children at all. When that becomes the focus of media attention, they have begun to get more worried. Most worrying is when issues on the environment remain unsolved and get worse by the day. There will be a time our youths might reach a stage when hope also disappears. A person can be worried but hopeful at the same time. But if hope is gone, it's a disaster because dreams and aspirations are also lost. They tend to do nothing but argue endlessly about it being too late to do anything.

Levels of worry differ around the world, depending on where you are. If you're in a place where there are a lot of weather-related events or if you're directly dependent on rainforest and mangroves, climate change can be worrying. People in cities may see things differently compared to indigenous people, farmers, or folks living in areas already experiencing the blights of global warming. Developing island nations, especially low-lying ones with coastal populations, are already constantly threatened by the effects of climate change. The island of Tuvalu is one of the low-lying islands where there is no point on the island higher than 4.5 metres above sea level. And the sea-level rise, there is estimated at around 1.2 millimetres each year. So by rough extrapolation, Tuvalu may eventually be inundated in at least fifty years in the future. I will be dead and gone by then, and why should I worry? We may be quick to react when environmental issues affect us directly, especially in matters that involve diseases and fatalities. We are also quicker and more serious in handling problems affecting the present generation. Issues that are bound to adversely affect future generations, the children of our children, we are often contented to leave the problems on the back burner. 'Let the future generations worry about that,' we may dismiss the problem. A touch of optimism might creep in by having an attitude, 'I am sure by then the latest advancement in climate science and technologies would provide timely answers for their safety and well-being.' Because of such prevailing mindset in us, we may be aware of future issues but tend to ignore them for the meantime.

But how quickly we forget. Mangroves have saved human lives in recent past. The tsunami that struck Indonesia on 26 December 2004 obliterated vast areas of Aceh Province on Sumatra Island, causing an estimated 230,000 deaths. As the tidal wave dissipated, some sixty thousand hectares of rice fields were left flooded with salt water and buried with sand. An estimated thirty thousand hectares of mangroves were destroyed. There was, however, an upside amidst this carnage that led to our realisation on the value of mangroves. As the powerful waves were lashing their devastating forces, the mangroves captured and dissipated some of the tsunami's energy. The mangroves acted as natural barriers that protected the villages and people living behind or within the proximity of the ecosystem. As little as 100 metres of dense mangroves could have reduced the destructive energy of a tsunami wave by as much as 90 per cent. That could have made the difference between life and death for tens of thousands of people in Aceh in 2004. Those living in many villages facing the sea without mangroves to cushion and mitigate the forces of the tsunami suffered the worst. Prior to that fateful calamity, huge areas of mangroves had been chopped down and ponds dug to farm shrimp and milkfish. Worldwide mangroves were being cleared at a rate of around 1 per cent a year, several times faster than the rate of deforestation on land. Let this be a bitter lesson. The past practice of converting mangrove swamps for economic development, aquaculture, industrial sites, and human settlement, stripping this barrier along large stretches of coastlines, needs to cease.

No living person today will ever forget the year 2020. It's the year Covid-19 caused horrendous carnage to our modern world. Its devastating impact has yet to be felt fully. All nations of the world were caught unaware, totally unprepared how to handle a pandemic that had the potential of changing the way we live in the future. Suddenly, the old familiar issues of deforestation and newly emerging diseases have attracted much attention. The role of reservoir hosts in the transmission of diseases has sparked much interest amongst scientists and conservationists. For decades now, our natural ecosystems have been rapidly altered by human activities. Pathogens, especially viruses, depend for their survival on the healthy functioning of the ecosystems because they are an integral part of the natural habitats. These can continue existing in the pristine forests where they were; after all, they have been in harmony with nature unnoticed with no adverse effects on human existence for millions of years. But this scenario is fast changing. As we continue to alter the natural ecosystems, we find ourselves plagued by newly emerging diseases capable of causing much havoc and many deaths. In recent times, the vast and rapid clearing of the Malaysian rainforests has resulted in the emergence of Nipah virus infections. Deforestation has caused the migration of the Malayan flying fox, the reservoir host of Nipah virus, from their original forest habitats to mango trees within the proximity of pig farms. They occurred as they searched for food sources, which were rapidly being depleted in their natural habitats. From there, they introduced their virus-laden saliva, urine, and excrement to the pigs, which in turn infected 257 people, killing 105 of them. This is hardly an isolated example. It is now well established that the virus causing HIV-AIDS, which currently infects more than thirty million people worldwide and has killed more than twenty-five million since 1981, was transmitted to human beings from the wild. People in West and Central Africa were exposed to the body fluids of infected chimpanzees that had intruded into human settlements in search of food.

Mangrove is a marvel of nature that is very awe-inspiring in many ways. I summarily mulled over what I have experienced in all my years of exploring this unique ecosystem. I reflected on what a mangrove forest meant to different people, in all its many aspects of ecological functions, natural resources, environmental services, education, wonderments, and tranquillity. I pondered over its future survival and continuation as the sea continued to eternally caress the sandy shores and muddy riverbanks, for this had been its only way of breathing and nourishing itself. Day in and day out, it would be drenched, inundated, and exposed to whatever debris from the land and the chemicals dumped by humans in the name of development and progress. Sadly, this has been humankind's way of expressing gratitude to the multitude of spectacular services and splendour of Mother Nature.

To the casual beachgoer, the sea by the mangrove forests affords an unending source of fresh breeze. To the morning beach joggers, the mangroves are the fountain of fresh oxygen and energy that overwhelm and fill their lungs and muscles for the day's activities. To the fishermen living in its proximity, the mangrove forest has been the sole provider, bountiful one day but gobbled up by stormy weather the next day. To the artistically inclined painters and shutterbugs, the mangroves and its inhabitants have been a wealth of inspiration for their creative works. But to me, it has been one vast, serene environment constantly accompanied by the eternal splatters and sprays of the waves. Subconsciously, there is something magical about this surging mass of water ebbing in and out of the mangroves. Time and again, its vibes and aura bring out the best of my cerebral endeavour. The mystique and grandeur of the mangroves have captivated mankind since the dawn of human history. Each of us has merely inherited this earth for the period of one's life. So the right thing to do would be to pass it intact to the next kith and kin. To abuse nature's bounty for our short-term gains would be an outright recklessness, a travesty of trust and responsibility to generations after us.

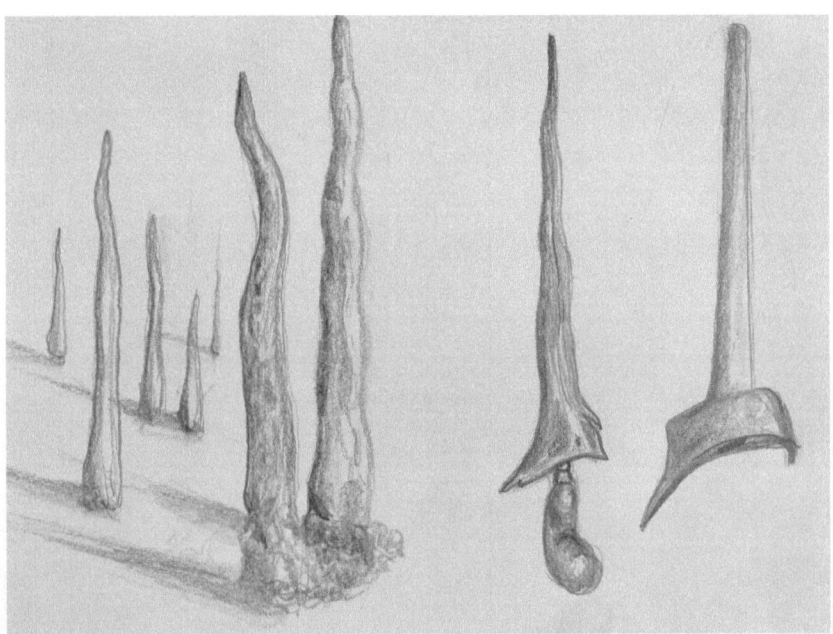

As a child, I used to don the aerial root of *Rhizophora* around my waist, pretending I was a fearsome Malay warrior walking around the village with a *kris*, a traditional dagger weapon of the Malay kingdom.

My childhood fascination with the mangroves started with seeing the aerial roots of *Rhizophora* sticking out of the mud, pointing straight to the sky. My friends and I used to don these dagger-shaped roots around our waists, pretending they were our deadly kris, the traditional weapon worn by Malay warriors. At seventy-two, I'm still wielding this imaginary kris in my head as a warrior, this time as an environmental warrior waging an educative war against enemies of mangroves, that magical ecosystem I was captivated and fell deeply in love with since I was eight years old.

Ghazally Ismail grew up in a small Malaysian village, Jelawat. Selected to attend a premier boarding school, the Malay College Kuala Kangsar, 400 km away from home, he missed his childhood environment amidst lush pristine rainforests, mangroves and sandy beaches. He earned his doctorate from the Indiana University School of Medicine, Indianapolis, USA in Immunology and Microbiology in 1976. For more than forty years, he spent his working life in Sabah and Sarawak on the island of Borneo.

ABOUT THE AUTHOR

As a professor, his research and teaching took him to exploring numerous rainforests and coral reefs of immense biodiversity. He published extensively not only in immunology and microbiology but also environmental issues pertaining to biodiversity and conservation. After retiring from the academia, he delved into creative activities including writing, drawing, photography and film documentaries. His previous books include The Malaysian Rainforest Realms: Fascinating Facts in Q&A (Marshal Cavendish 2010), Nature's Pearls of Wisdom (UKM Publisher 2013), Monkey Moments (Xlibris Publication 2021). He was the executive producer of nature film documentaries "The Guardian of Kinabalu" and " The Giant's Guardian" about the world biggest flower Rafflesia.

www.ingramcontent.com/pod-product-compliance
Lightning Source LLC
Chambersburg PA
CBHW021401210526
45463CB00001B/186